Bifurcations in Flow Patterns

Nonlinear Topics in the Mathematical Sciences

An International Book Series dealing with Past, Current and Future Advances and Developments in the Mathematics of Nonlinear Science

---

Editor:

VOLUME 2

Bifurcations in Flow Patterns

Some Applications of the Qualitative Theory of Differential Equations in Fluid Dynamics

by

P. G. Bakker
Department of Aerospace Engineering,
Delft University of Technology,
Delft, The Netherlands

KLUWER ACADEMIC PUBLISHERS
DORDRECHT / BOSTON / LONDON

Library of Congress Cataloging-in-Publication Data

Bakker, P. G.
Bifurcations in flow patterns : some applications of the qualitative theory of differential equations in fluid dynamics / by P.G. Bakker.
p. cm. -- (Nonlinear topics in the mathematical sciences ; v. 2)
Includes bibliographical references and index.
ISBN 0-7923-1428-X (acid-free)
1. Fluid dynamics. 2. Differential equations--Numerical solutions. 3. Bifurcation theory. I. Title. II. Series.
QA911.B255 1991
532'.05'0151535--dc20 91-29830

ISBN 0-7923-1428-X

Published by Kluwer Academic Publishers,
P.O. Box 17, 3300 AA Dordrecht, The Netherlands.

Kluwer Academic Publishers incorporates
the publishing programmes of
D. Reidel, Martinus Nijhoff, Dr W. Junk and MTP Press.

Sold and distributed in the U.S.A. and Canada
by Kluwer Academic Publishers,
101 Philip Drive, Norwell, MA 02061, U.S.A.

In all other countries, sold and distributed
by Kluwer Academic Publishers Group,
P.O. Box 322, 3300 AH Dordrecht, The Netherlands.

Printed on acid-free paper

Printed in the Netherlands

CONTENTS

PREFACE

The main idea of the present study is to demonstrate that the qualitative theory of differential equations, when applied to problems in fluid- and gasdynamics, will contribute to the understanding of qualitative aspects of fluid flows, in particular those concerned with geometrical properties of flow fields such as shape and stability of its streamline patterns.

It is obvious that insight into the qualitative structure of flow fields is of great importance and appears as an ultimate aim of flow research. Qualitative insight fashions our knowledge and serves as a good guide for further quantitative investigations. Moreover, qualitative information can become very useful, especially when it is applied in close correspondence with numerical methods, in order to interpret and value numerical results. A qualitative analysis may be crucial for the investigation of the flow in the neighbourhood of singularities where a numerical method is not reliable anymore due to discretisation errors being unacceptable.

Up till now, familiar research methods - frequently based on rigorous analyses, careful numerical procedures and sophisticated experimental techniques - have increased considerably our qualitative knowledge of flows, albeit that the information is often obtained indirectly by a process of a careful but cumbersome examination of quantitative data.

In the past decade, new methods are under development that yield the qualitative information more directly.

These methods, make use of the knowledge available in the qualitative theory of differential equations and in the theory of bifurcations.

The qualitative theory of differential equations as applied to dynamical systems of two and three-dimensional vector fields appears to be very useful, in order to determine the topology and structural stability of streamline patterns, occurring in fluid dynamics. The theory originates from the work of Poincaré and is further developed by Birkhoff, Lyapunov, Andronov and his co-workers, Arnold and many others.

This theory is applied to flows, which can be described by two variables x and y and the streamlines are the solution curves of a system of the second-order, $\dot{x}=P(x,y)$, $\dot{y}=Q(x,y)$

A survey of some elements of the qualitative theory of differential equations is given in Chapter I. The review introduces and discusses in some detail the most important concepts, theorems, and methods that are used in the subsequent analyses; for rigorous mathematical proofs the reader will be referred to the literature. Actually Chapter I may be consulted in order to be informed about notions as phase portraits of a dynamical system, singular points, topological structure, structural stability, degenerate systems, bifurcation of vector fields, unfoldings, co-dimension and center-manifold theory.

The first application of the qualitative theory concerns the flow geometry of three-dimensional inviscid conical gas flows. It is treated in Chapter II. Conical flows have the specific property that the velocity of the gas particles and the quantities, defining the state of the gas, e.g. pressure, density and temperature, are constant along rays originating from a common point (conical center). A three-dimensional conical flow is essentially two-dimensional and is decribed adequately on a unit sphere (around the conical center) on which the flow geometry is displayed by conical streamlines. Conical streamlines are defined as follows. For conical flows, the family of spatial streamlines passing the same ray form a conical stream surface with vertex in the conical center. The intersection of a conical stream surface with the unit sphere is a conical streamline. Along a conical streamline the entropy is constant or it jumps if a shock is passed.
The conical flow geometry on the unit sphere is governed by a second-order dynamical system. The corresponding vector field is determined by the cross-flow velocity, being the velocity component tangential to the unit sphere. In points on the unit sphere where this component vanishes the flow is purely radial; these points are called conical stagnation points and appear as singularities of the second-order dynamical system governing the conical streamlines. In Chapter II, possible flow patterns near conical stagnation points are studied and their structural stability is examined.
Transitions to a different conical flow pattern are interpreted as bifurcations of structurally unstable higher order conical stagnation points.
Chapter II gives a complete classification of first order conical stagnation points and starts with the classification of higher order conical stagnation points, by considering second- and third-order points, these points being of most practical interest in real flow problems.
Chapter II ends up with a qualitative examination of the inviscid flow around three different conical bodies: circular cones, delta-wings with arrow-shaped cross-section and external corners.

In Chapter III the qualitative theory will be applied to steady two-dimensional incompressible viscous flows along the surface of a plane or along a slightly curved wall. A second-order dynamical system, whose trajectories represent streamlines, is derived.
A singular-point analysis is performed in order to obtain detailed information about the local flow topology.
In particular, singular points on the wall surface are of interest, because in these points the shear stress vanishes indicating flow separation or flow attachment.

As in Chapter II the singular points are distinguished into first-order and higher-order singularities, the former appear only as centerpoints and saddle points. First-order saddle points that are located at the wall surface represent solutions that are recognized as the classical Oswatitsch-Legendre solution for flow separation or attachment. The Oswatitsch-Legendre solution is structurally stable in the sense that analytical perturbations in streamwise pressure gradients and/or shear stress gradients will not affect the flow topology near the separation (attachment) point.
Apart from the structurally stable solutions, structurally unstable solutions appear as well, they occur as higher-order or degenerate singularities of the dynamical system that governs the streamline pattern.
The study of these degenerate singularities; their unfoldings and bifurcational behaviour, will be the main subject of Chapter III. Unfoldings, resulting in degenerate singularities falling apart into a number of first-order singularities, describe possible structurally stable flow patterns in viscous incompressible flow. These patterns, being more complex for singularities with a higher degree of degeneracy, are interpreted physically. Some of them appear to be very common in aerodynamics, others are new and are concerned with topics as:
- genesis of laminar separation bubbles
- flow separation on moving walls (Moore-Roth-Sears criterion)
- interference of separations and attachments, and
- formation of asymmetric standing eddies in the near wake behind a body.

I am indebted to Willem Bannink and John Reyn for their useful suggestions and lively discussions which enables me to improve the content and style of the manuscript.
Moreover I wish to thank Jan Boeker for his careful reading of the written material, especially his many corrections to it have been helpful.
Furthermore, I am grateful to the Aerospace Department of the Delft University of Technology that provided me with all facilities for reproduction of text and figures.
I wish to mention the careful typing of Marian van Laren and Marijke Mooyman, they did an excellent job.
Finally I am very grateful to Willem Spee, who transformed the rather imperfect sketching into significant pictorial material.

June 1991 P.G. Bakker

CHAPTER I. SOME ELEMENTS OF THE QUALITATIVE THEORY OF DIFFERENTIAL EQUATIONS

1. Phase space representation of a dynamical system

With the aim of easy reference we will give in this chapter some elements of the qualitative theory of dynamical systems which we will use in the next chapters.
The theory of dynamical systems has been extensively studied over a long period of time. Although many questions remain as yet unsolved, a large amount of results has been obtained and is available for applications.
Numerous textbooks, articles and papers testify of the progress made in this branch of mathematics.
A comprehensive survey of important achievements, including recent developments and advanced methods, is given by Guckenheimer and Holmes (1983): 'Nonlinear Oscillations, Dynamical Systems and Bifurcations of Vector Fields'.
This book, written with a strong view to combine pure mathematical reasoning and applicability to practice, has inspired me in writing this chapter.

The theory of dynamical systems aims to study the time behaviour of evolutionary systems which are described mathematically by an equation of the form:

$$\frac{dX}{dt} = \dot{X} = f(X,t)$$

where $X = X(t) \in \mathbf{R}^n$ is a n-vector and f(X,t) is a sufficiently smooth function defined on some subset $U \subseteq \mathbf{R}^n \times \mathbf{R}$.
In applications t is usually interpreted as time. The function f defines a vector field in the n-dimensional space the so-called *phase space*.
The solutions of the system, representing sequential states of the evolutionary process, appear as integral curves of the vector field in the phase space.
In the literature on this subject, one encounters several names for these curves, we quote: solution curves, trajectories, (phase-)paths and orbits. In the same way, the terms phase portrait, phase pattern, flow or trajectory pattern are used to indicate the whole set of solution curves in the phase space.

Dynamical systems where f does not explicitly depend on time are called *autonomous* and the trajectories of such systems do not change if time goes on.
In the following we only need to study these autonomous systems, being expressed by the equation

$$\dot{X} = f(X), \quad X \in \mathbf{R}^n \tag{1.1}$$

In a classical treatment of such a system, see for example Coddington and Levinson (1955), the attention is mainly directed to the properties of *individual* solution curves and culminates into questions about explicit time behaviour and dependency on initial conditions of the solutions.
A different approach to the study of dynamical systems is obtained if one considers families of solution curves; then qualitative questions arise such as: do there exist steady and/or periodic solutions, what are their stability properties.

Domains in the phase space where f(X) is a nonvanishing vector function are called *regular*; the phase portraits in a regular domain are relatively simple as they can be mapped by a homeomorphism onto a family of parallel trajectories, see Fig. 1.1.

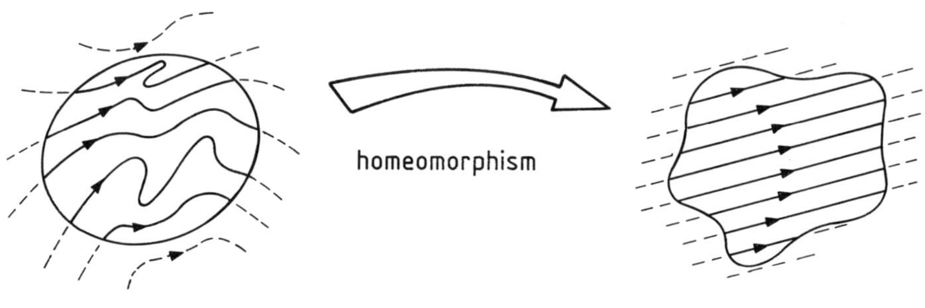

Fig. 1.1. Mapping of the paths in a regular domain into a field of parallel paths.

An important class of solutions of Eq. (1.1) are the so-called *equilibrium* or *steady solutions*: $X = X_o$ with X_o satisfying $f(X_o) = 0$. An equilibrium or steady state solution is thus represented in the phase space by a single point and corresponds to a critical point of the vector field, in the following such a point will often be referred to as a singular point of the differential equation (1.1). Steady solutions: $X = X_o$ cannot be reached by a neighbou-

ring solution X(t) in finite time. A steady solution is *stable* if a solution X(t) through a point in a neighbourhood of X_o, remains close to that steady solution for $t \rightarrow \infty$. The steady solution is asymptotically stable if all neighbouring solutions $X(t) \rightarrow X_o$ as $t \rightarrow \infty$.
Steady solutions not satisfying the stability condition are said to be unstable.
Some examples of stable and unstable solutions in the phase space are depicted in Fig. 1.2.

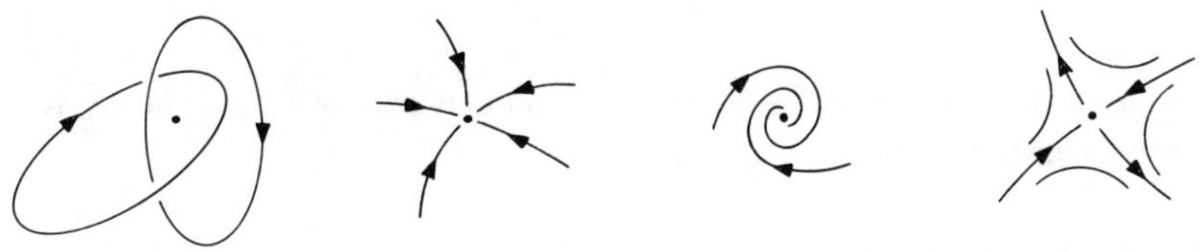

Fig. 1.2. Stable and unstable steady solutions.

The phase portrait of trajectories near singular points can be rather complicated and is usually not homeomorphic with a field of parallel trajectories.
Important elements of phase portraits of systems are singular points and the local trajectory pattern near these points.
Therefore, a systematic description of phase portraits near singular points is a useful tool when analysing phase portraits of systems. A detailed treatment of phase patterns near singular points is given in paragraph 2 of this chapter.

Another class of solutions of Eq. (1.1) we want to mention here are the so-called periodic solutions satisfying: X(t+T) = X(t) with period T. Periodic solutions of a dynamical system appear as closed paths in the phase space.
Closed paths in $\mathbf{R}^2$, representing periodic solutions of two-dimensional systems, must divide the phase plane into an inner and an outer region. If there exists a neighbourhood of a closed path which does not contain another closed path, this path is called a *limit cycle.*
A limit cycle is stable if all neighbouring paths of the limit cycle approach it if $t \rightarrow \infty$, otherwise the limit cycle is unstable: see Fig. 1.3.
An important question, but often difficult to answer, is to determine whether an autonomous system has periodic solutions and where they appear in the phase space.

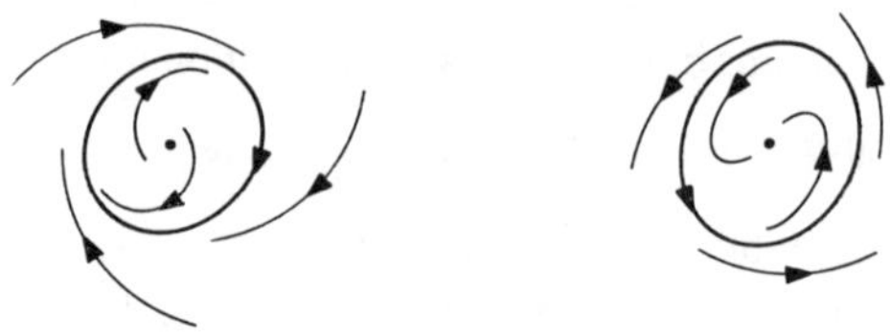

Fig. 1.3. Limit cycle in R^3

An elementary condition that can be used for the non-existence of closed paths is formulated by Bendixson's criterion.

Bendixson's criterion

Let $\dot{x} = P(x,y)$ and $\dot{y} = Q(x,y)$ be an analytical dynamical system and let U be a simply-connected domain of the phase plane on which the divergence of the vectorfield: div(P,Q) does not change sign and is not identically zero. Then there are no closed paths lying entirely in U.

The proof of Bendixson's theorem goes as follows.

Application of the divergence theorem along a closed curve γ lying entirely in U gives

$$\int_{U_\gamma}\int \left[\frac{\partial P}{\partial x} + \frac{\partial Q}{\partial y}\right] dx\, dy = \int_\gamma (P,Q) \cdot \underline{n}\, ds$$

With U_γ the interior of γ, $\underline{n}$ the outward normal and ds a line element of γ.

If γ is a path of the vector field (P,Q), then (P,Q) and $\underline{n}$ are perpendicular so that the line integral $\int_\gamma$ vanishes identically. But since the integrand of the integral $\int_{U_\gamma}\int$ is of one sign, the integral cannot be zero. Then the curve γ cannot be a path of the vector field (P,Q).

2. Phase portraits near singular points

Assume that a dynamical system $\dot{X} = f(X)$, $X \in \mathbf{R}^n$ has a singular point X_o so that $f(X_o) = 0$. In order to characterize the trajectory pattern near X_o we assume $f(X)$ to be sufficiently smooth and we expand $f(X)$ near X_o. Retaining only the linear part in the expansion there follows with $X - X_o = \xi$

$$\dot{\xi} = Df(X_o)\,\xi, \quad \xi \in \mathbf{R}^n \tag{1.2}$$

where $Df(X_o)$ denotes the Jacobian matrix $[\frac{\partial fi}{\partial xj}]$ of the first order partial derivatives of the vector function

$$f = \begin{bmatrix} f_1(x_1, \ldots x_n) \\ f_2(x_1, \ldots x_n) \\ \cdot \\ \cdot \\ \cdot \\ f_n(x_1, \ldots x_n) \end{bmatrix}$$

evaluated in the singular point X_o.

Equation (1.2) is a linear system with constant coefficients and can be analysed with classical methods yielding the corresponding 'linearized' phase portraits. However these, 'linearized' phase portraits are not necessarily equivalent with those near X_o of the original non-linear system.

The relation between both phase portraits is given by the theorem of Hartman-Grobman which holds for systems in $\mathbf{R}^n$.

Hartman-Grobman

If $Df(X_o)$ has no eigenvalues with zero real part, then the family of trajectories near a singular point X_o of a nonlinear system $\dot{X} = f(X)$ and those of the locally linearized system have the same topological structure; which means that in a neighbourhood of X_o there exists a homeomorphic mapping which maps trajectories of the non-linear system into trajectories of the linear system.

When $Df(X_o)$ has no eigenvalues with zero real part, the singular point X_o is called *hyperbolic* or *nondegenerate.*

The theorem of Hartman-Grobman implies that local linearization near singular points will be an effective method when analysing phase portraits of non-linear systems. Therefore it is worthwhile to recall first the theory of linear systems in some detail.

Linear systems

Linear systems in $\mathbf{R}^n$ are treated extensively and in various aspects in the literature. Especially, systems in $\mathbf{R}^2$ are very well known and receive much attention in many textbooks on ordinary differential equations, e.g. Coddington and Levinson (1955) and Jordan and Smith (1977).
Furthermore, for linear systems in $\mathbf{R}^3$ reference may be made to Reyn (1964), where a sound treatment of singular points culminates into a conveniently arranged survey of three-dimensional phase portraits.

Let us briefly recall here the main results for linear systems in $\mathbf{R}^2$, as they are frequently used in the following chapters.
Consider the two-dimensional system

$$\dot{X} = AX$$

where A is a constant 2x2 matrix and $X = (x,y)^T$.
The system has only one singular point: the origin (0,0).
A classification of phase portraits can be given in terms of the trace: $p = \lambda_1 + \lambda_2$ and Jacobian: $q = \lambda_1 . \lambda_2$ of A; λ_1, λ_2 being the eigenvalues of A.
We consider several cases

(i) $\mathbf{q < 0}$: the eigenvalues λ_1 and λ_2 are real with different signs.
The singular point is a saddle point and is unstable.

(ii) $\mathbf{0 < q < (\frac{1}{2}\,p)^2}$: the eigenvalues λ_1 and λ_2 are real, unequal and of the same sign. The singular point is a stable node (sink) if $p < 0$ and an unstable node (source) if $p > 0$.

(iii) $\mathbf{q > (\frac{1}{2}\,p)^2}$ **($\mathbf{p \neq 0}$)**: the eigenvalues λ_1 and λ_2 are conjugate complex with non zero real part. The singular point is a stable focus if $p < 0$ and an unstable focus if $p > 0$.

(iv) **$\mathbf{q > 0,\ p = 0}$**: the eigenvalues λ_1 and λ_2 are purely imaginary. The trajectories form closed curves surrounding the origin and the singular point is called a center.

(v) $\mathbf{q = (\frac{1}{2}\,p)^2}$ **($\mathbf{p \neq 0}$)**: the eigenvalues λ_1 and λ_2 are real and equal ($\lambda_1 = \lambda_2 = \lambda$). The nature of the phase portrait depends on the Jordan form of A. If A has the Jordan form $\begin{bmatrix} \lambda & 0 \\ 0 & \lambda \end{bmatrix}$ the trajectories form a star shaped node, which is stable for $p < 0$ and unstable if $p > 0$.
If A has the Jordan form $\begin{bmatrix} \lambda & 1 \\ 0 & \lambda \end{bmatrix}$ the phase portrait is called an inflected node; all trajectories tend to the origin in the same direction and are parallel at infinity.

(vi) **$\mathbf{q = 0,\ p \neq 0}$**: one of the eigenvalues is zero; the phase portrait consists of a family of parallel paths and a line of singular points, which is stable for $p < 0$ and unstable if $p > 0$.

This classification of linear systems is conveniently arranged in the p-q-plane, shown in Fig. 1.4.

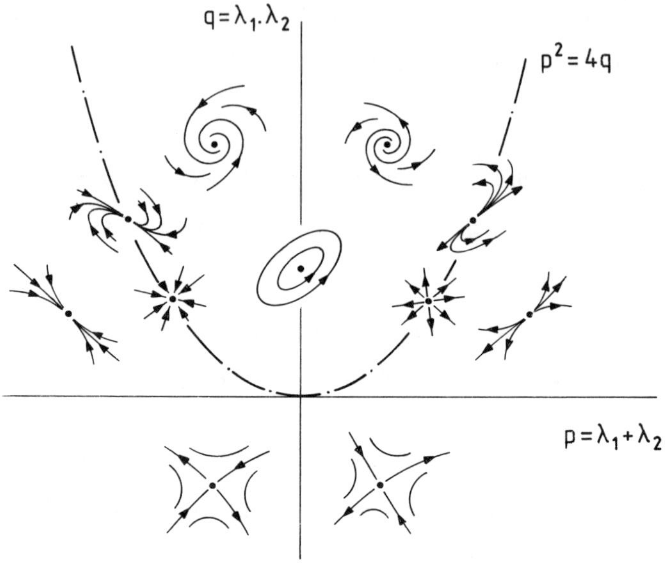

Fig. 1.4. General classification of a linear system in R^2.

The isolated singular points for a linear system are saddles, nodes, foci and centers. Since nodes, foci and saddle points satisfy $\text{Re}(\lambda) \neq 0$, Hartman-Grobman's theorem implies that adding sufficiently smooth non-linearities does not change the phase portraits near these types of singular points.
Centers in non-linear systems satisfy $\text{Re}(\lambda) = 0$, and their existence cannot be shown by linearization. As an illustration consider the nonlinear system

$$\dot{x} = y$$

$$\dot{y} = -x + \varepsilon x^2 y$$

with eigenvalues $\lambda_1, \lambda_2 = \pm i$. Unless $\varepsilon = 0$, the singular point (0,0) is not a center as in the linearized system, but a nonlinear focus, stable if $\varepsilon < 0$.

Nonhyperbolic singularities as characterized by a vanishing Jacobian (q = 0) can appear in the phase space in different forms: isolated points, curves etc. The isolated points are usually denoted as *multiple-equilibrium* points, *degenerate* singularities or *higher-order* singularities. The topological structure of the local phase portrait near higher-order singularities can be very complicated and a general classification for singularities in $\mathbf{R}^n$ is hard to give.

For $\mathbf{R}^2$, Andronov et. al (1973) offer a classification for isolated higher-order singularities having a nonvanishing degenerate linear part. These points will be reviewed in more detail in paragraph 4.

3. Topological structure of phase portraits, structural stability, bifurcation

3.1 Topological structure of phase portraits

The concept of topological structure of phase portraits has already been introduced in paragraph 1 when we mentioned the fundamental theorem of Hartman-Grobman.

Now a more thorough examination of this concept will be given because it enables us to introduce and explain the concept of structural stability.

Characteristic features of the phase portrait which may be called qualitative properties are for example the number and type of singular points, the existence of closed paths and regions of attraction. Formally, one may define qualitative properties as those properties of the phase portrait which remain invariant under a topological mapping or homeomorphism. A topological mapping between two regions in de plane is a one-to-one and bicontinuous mapping, meaning that each point M is mapped exactly onto one point M' and that distinct points M_1 and M_2 are mapped onto distinct points M_1' and M_2', and the mapping is continuous either way.

An intuitive description of a topological mapping of the plane onto itself is given by Andronov et. al (1973) as follows:

> ***'imagine the plane is to be made from rubber which is deformed in some way, stretching and squeezing it at various points, but without tearing or folding. Any topological mapping of the plane into itself is either a deformation of the above type (without tearing and folding) or a mirror-reflection of the plane followed by such a deformation'.***

It may be clear that a topological mapping of a phase portrait can result in drastic changes in the shape of trajectories, but certain properties will nevertheless be preserved. As an example: singular points are mapped into singular points and closed curves remain closed curves.

The concept of topological structure of a phase portrait in $\mathbf{R}^2$ is now indirectly introduced by the following definition.
Consider in a region $G \subset \mathbf{R}^2$ two dynamical systems D_1 and D_2, given by

$$\dot{X} = f_1(X) \qquad (D_1)$$
$$\dot{X} = f_2(X) \qquad (D_2)$$

The phase portraits of the systems (D_1) and (D_2) have the same topological structure if there exists a topological mapping (homeomorphism): T which maps G onto G and which takes paths of (D_1) over into paths of (D_2). If the topological mapping T (with inverse T^{-1}) is k times differentiable (k> 0), the mapping is called C^k diffeomorph and the two vector fields are said to be C^k-equivalent; C^k-equivalence with $k > 0$ implies that certain smoothness properties (k-times differentiable) of trajectories remain preserved by the mapping process.
If two points X_1 and X_2 lie on the same path of system D_1, then their images TX_1 and TX_2 lie on the same path of system (D_2).

Also, if two points X_1 and X_2 lie on the same path of system (D_2), then their images $T^{-1}X_1$ and $T^{-1}X_2$ lie on the same path of system (D_1).
The given definition of topological structure is in a certain sense indirect since it does not state exactly what topological structure is, but it specifies the necessary conditions for equal topological structures.

3.2 Structural stability; bifurcation

After the definition of equivalent topological structure, we now introduce the concept of structural stability.

Consider the dynamical system (S) defined in a region $G \subset \mathbf{R}^n$.

$$\dot{X} = f(X) \qquad (S)$$

System (S) is said to be *structurally stable* if an infinitesimal change of f(X) leaves the topological structure unaffected in G, otherwise the system is called structurally unstable.

It should be noticed, that structural stability is not an intrinsic property of a topological structure but is related to the class of infinitesimal changes of f(X) that are allowed. Both the class of perturbations of the vector field which are admitted and f(X) itself determine whether there is structural stability or not.

This characteristic property of structural stability is illustrated in the next example where two different types of perturbations are imposed on the linear system (in $\mathbf{R}^2$):

$$\begin{aligned} \dot{x} &= -x \\ \dot{y} &= -2y \end{aligned} \tag{1.3}$$

which has a node at the origin.
First we consider an analytic perturbation by adding a linear term μx to the right-hand side of $\dot{y}$:

$$\begin{aligned} \dot{x} &= -x \\ \dot{y} &= -2y + \mu x \end{aligned} \tag{1.4}$$

with μ being a small perturbation parameter.

The origin is a stable node for $\mu = 0$ as well for $\mu \neq 0$, thus (1.3) is structurally stable with respect to the perturbations of the (analytical) character given in Eq. (1.4).

Next we consider a non-analytic perturbation of (1.3) by adding a perturbation of the form $\mu \sqrt{|x|}$

$$\begin{aligned} \dot{x} &= \mu \sqrt{|x|} - x \\ \dot{y} &= -2y \end{aligned} \tag{1.5}$$

If $\mu > 0$ system (1.5) has two singular points on the x-axis: (0,0) and $(\mu^2,0)$. Near (0,0) the trajectories behave partially like those near a stable node and partially like those near a saddle point. At $(\mu^2,0)$ there occurs a stable node. For $\mu < 0$ similar results will follow. The phase portraits obtained if μ varies near zero are shown in Fig. 1.5.

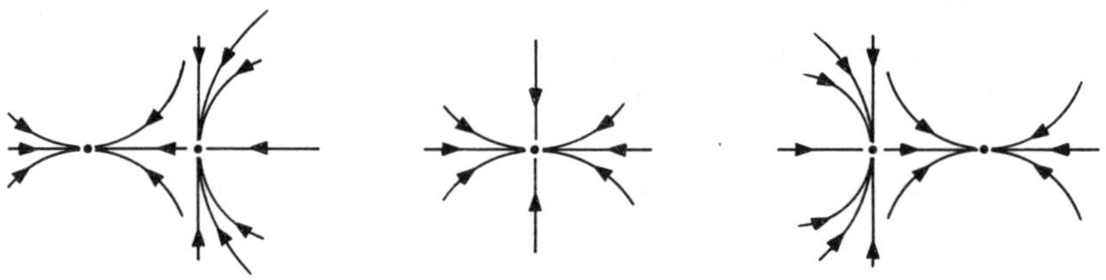

Fig. 1.5. Nonanalytic perturbation of a nodal point.

The examples illustrate that the stable node at (0,0) is structurally stable against analytical perturbations (Eq. (1.4)) but structurally unstable against perturbations (nonanalytical) as given by Eq. (1.5). Evidently, structural stability of a topological structure cannot be established independent of the class of perturbations that are imposed on the dynamical system.

To see more aspects about structural stability and leaving the discussions as simple as possible, let us continue by considering systems in $\mathbf{R}^2$ in some more detail. Assume that the linear system:

$$\dot{X} = A\,X \qquad X \in \mathbf{R}^2 \tag{1.6}$$

is perturbed by adding a term $\mu\, f(X)$ yielding the nonlinear system

$$\dot{X} = A\,X + \mu\, f(X) \qquad \mu \in \mathbf{R} \tag{1.7}$$

where f(X) is at least C^1 and f(0) is not necessarily equal to zero.
If the eigenvalues λ_1, λ_2 of A have non-vanishing real parts, $Re(\lambda_{1,2}) \neq 0$ then it can be shown that near the origin the phase portrait of (1.6) is structurally stable against the C^1-

perturbations given in Eq. (1.7). To this end, the phase pattern of Eq. (1.7) near the origin has to be determined. A singular point: X_o of Eq. (1.7) satisfies

$$A X_o + \mu f(X_o) = 0$$

Since A is an invertible matrix, and $X_o = -\mu A^{-1} f(X_o)$ the implicit function theorem can be used to find near X = 0 the unique solution $X_o = O(\mu)$ for sufficiently small μ.
The phase portrait of Eq. (1.7) follows by considering the locally linearized system near X_o, of which the coefficient matrix $A + \mu Df(X_o)$ has eigenvalues which depend continuously on the perturbation parameter μ. For small μ, compared with $|Re(\lambda_{1,2})|$, ($Re(\lambda_{1,2}) \neq 0$) the eigenvalues of $A + \mu Df(X_o)$ cannot cross the imaginary axis so that X_o is also a hyperbolic singularity. From Hartman-Grobman's theorem, then follows that the systems (1.6) and (1.7) have phase portraits near (0,0) which are topologically equivalent.

Hence, phase portraits near hyperbolic points are structurally stable with respect to C^1-perturbations, phase portraits near nonhyperbolic points may loose their topological structure if C^1-perturbations are imposed on the dynamical system. This phenomenon where a small variation of the system causes a change of topological structure is called a *bifurcation.* If the changes are caused by perturbations containing parameters, those parameters which actually cause bifurcation are called *bifurcation parameters.*

Two types of bifurcations in phase portraits may be distinguished: *local* bifurcations and *global* bifurcations.
The former appear if the topology of the phase portrait is only locally affected. Such bifurcations can be observed and described with local analyses, they occur in particular if nonhyperbolic points are present; bifurcation changes the topology of the phase portrait only in a small neighbourhood of the nonhyperbolic singularity. In those cases where local analyses fail to detect bifurcation effects, the bifurcation is called a global bifurcation. Saddle-connections and multiple limit cycles are well known examples of global bifurcations. Let us discuss them briefly for phase portraits in $\mathbf{R}^2$.

3.3 Saddle connections

Consider a dynamical system in $\mathbf{R}^2$ having a phase portrait which contains a special trajectory connecting two (hyperbolic) saddle points. Suppose that this system is perturbed so that the saddle connection cannot be maintained. Although the topology in the phase plane is significantly altered, see Fig. 1.6, the disappearance of the saddle connection cannot be detected by only a local examination of the phase portraits; to observe the broken saddle connection a broader view, containing the saddle points together with the separatrices, is necessary. The imposed perturbations have a so-called global effect on the trajectory pattern and as a result the transition process is called a global bifurcation.

Fig. 1.6. Perturbation of saddle connection, global bifurcation.

3.4 Multiple limit cycle

Another example of a global bifurcation may appear if an isolated closed path: γ_c is embedded in a trajectory pattern in $\mathbf{R}^2$ as shown in Fig. 1.7. Assume that for increasing t the trajectories in the outer region are spiralling inwards, all tending to γ_c if $t \to \infty$. In the inner region of γ_c they tend to γ_c for $t \to -\infty$ and to a focal point for $t \to \infty$. The isolated closed path γ_c is a (semistable) *limit cycle*, which cannot be reached in finite time along trajectories starting in a neighbourhood of γ_c.

The phase portrait as sketched in Fig. 1.7 can be generated by the paths of the system

$$\frac{dr}{dt} = -r(r-1)^2 \quad , \qquad \frac{d\theta}{dt} = 1$$

where (r,θ) are polar coordinates in the phase plane. The closed path γ_c is then found on the circle $r=1$.

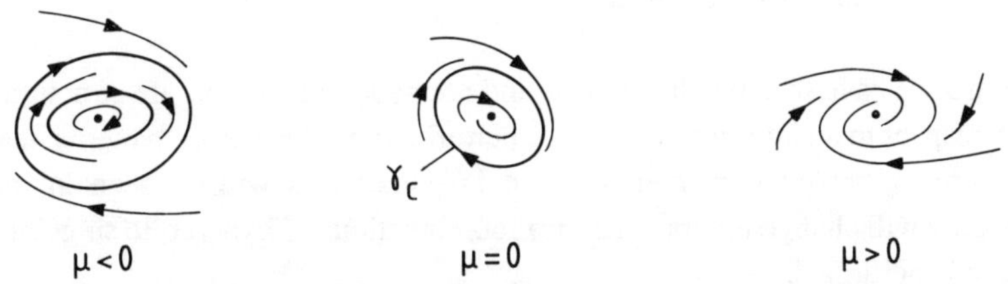

Fig. 1.7. Perturbation of limit cycle, global bifurcation.

Consider a perturbation of the system by adding the linear term μr ($\mu \in \mathbf{R}$) to the right-hand side of the equation $\frac{dr}{dt}$; μ appears to be a bifurcation parameter.
The phase portrait of the perturbed system

$$\frac{dr}{dt} = -r(r-1)^2 + \mu r \ , \qquad \frac{d\theta}{dt} = 1$$

has two closed paths: $r_{1,2} = 1 \pm \sqrt{-\mu}$ if $\mu < 0$ and no closed path if $\mu > 0$. The appearance (disappearance) of closed paths has a global effect on the topological structure of the phase portraits, it can not be established by only a local examination of the trajectory pattern.

The previous remarks point out that bifurcations - local or global - may appear if the phase portrait is structural unstable.
Nonhyperbolic singular points appear as elements in phase portraits that may cause local bifurcations; on the other hand saddle-connections and closed paths can give rise to bifurcations with a global character.
Whether a bifurcation actually occurs is not only determined by the presence of structural unstable elements in a phase portrait but depends also on the class of perturbations that is admitted.
The last statement applies also to global bifurcations, as can be shown by perturbing a Hamiltonian system: $\dot{x} = \frac{\partial H}{\partial y}$, $\dot{y} = -\frac{\partial H}{\partial x}$ such that the perturbed system remains Hamiltonian. Then closed trajectories can remain closed preventing a global bifurcation to appear.

4. Higher-order singularities in $\mathbf{R}^2$

In paragraph 3 we have seen that hyperbolic and nonhyperbolic points are structurally unstable with respect to an appropriate class of perturbations. Trajectory patterns near nonhyperbolic points can be very complex and in general, as will be seen in the next paragraph, they will change into patterns near combinations of hyperbolic singular points if the system is perturbed.
From this point of view one might argue that nonhyperbolic points are far from interesting because they are so 'exotic' that they are hardly met in practical situations where small perturbations, always present, lead them desintegrate.

However, nonhyperbolic points can be very helpful to analyse transition processes in phase portraits, where an appropriate change of parameters causes a change of the topological structure.
In that case, special combinations of parameter values exist at which one or more singular points in the phase pattern become nonhyperbolic. For such a parameter combination the system is called *degenerate* or *nongeneric*; however, a small change of these parameter values can convert the nongeneric system into a *non-degenerate* or *generic* one.
Such a degenerate state of the system marks a point in the parameter space at which the topological properties of the system change qualitatively.
These observations give the motivation for a more detailed treatment of these 'exotic' nonhyperbolic points.

The discussion will be restricted to nonhyperbolic points in $\mathbf{R}^2$ since only these will be found to occur in the subsequent part of this work.
The theory of nonhyperbolic singular points in $\mathbf{R}^2$ is well developed by Andronov et. al (1973) and the following review of the most important results serves as a basis for applications. For a profound treatment including mathematical proofs, we refer to Andronov et. al (1973).
Andronov considers an isolated nonhyperbolic singular point of an analytic vector field in $\mathbf{R}^2$ such that the expansion near the singular point involves at least one first-order term. Then a distinction can be made between singularities having one or both eigenvalues zero.

4.1 Higher-order singular points with one zero eigenvalue; $p \neq 0$

Suppose that O(0,0) is an isolated singular point of a planar system with one nonzero eigenvalue.
Then the system can be written as

$$\dot{x} = P(x,y) = ax + by + P_2(x,y)$$

$$\dot{y} = Q(x,y) = cx + dy + Q_2(x,y)$$

where $P_2(x,y)$ and $Q_2(x,y)$ are analytic with terms not lower than second degree and for the eigenvalues in O(0,0) we have

$$p = \lambda_1 + \lambda_2 = a + d \neq 0$$

$$q = \lambda_1 \lambda_2 = ad - bc = 0$$

This system can be transformed by a nonsingular linear coordinate transformation in the canonical form

$$\dot{x} = P(x,y) = P_2(x,y)$$

$$\dot{y} = Q(x,y) = y + Q_2(x,y)$$

For isolated points it is assumed that $P_2(x,y) \neq 0$.

For this system Andronov has shown that trajectories exist that tend to the singular point in a definite direction (semipaths). These semipaths tend to O(0,0) only in the directions 0, $\pi/2$, π and $3\pi/2$; only one semipath exists in the direction $\pi/2$ and only one in the direction $3\pi/2$; these two special semipaths are denoted L_1 and L_2 respectively.

To obtain the possible topological structures near the singular point consider the equation $y+Q_2(x,y)= 0$. By the implicit function theorem this equation has exactly one solution $y = \varphi(x)$ in a neighbourhood of O(0,0), where $\varphi(x)$ is analytic and obeys the condition $\varphi(o) = \dot{\varphi}(o) = 0$. The curve $y= \varphi(x)$ is an isocline of horizontal directions ($\dot{y} = 0$) of the vector field.

The next step is to define a function:

$$\psi(x) = P_2(x, \varphi(x))$$

which gives the value of $\dot{x}$ on the curve $\dot{y} = 0$. Because O(0,0) is isolated ($P_2 \neq 0$), $\psi(x)$ assumes the form

$$\psi(x) = \Delta_m x^m + o(x^m),\ m \geq 2$$

where Δ_m is the first nonvanishing coefficient ($\Delta_m \neq 0$) and m is integer.

Let us consider the neighbourhood U_δ of O(0,0) bounded by a circle C with radius δ, see fig. 1.8.

The curve $y = \varphi(x)$ divides the domain U_δ into two parts. Since $\dot{y} = y + Q_2(x,y)$, $\dot{y} > 0$ in the upper region, so that the vector field points upwards there.

Similarly we have downwash in the lower region. This implies that the semipaths L_1 and L_2, lying above and below the curve $y = \varphi(x)$ respectively, are unstable. Since $\varphi(x)$ is analytic, either $y = \varphi(x)$ coincides with the x-axis ($\varphi(x)=0$) or it has, in a sufficiently small neighbourhood of O(0,0), no other points than O(0,0) in common with the x-axis.

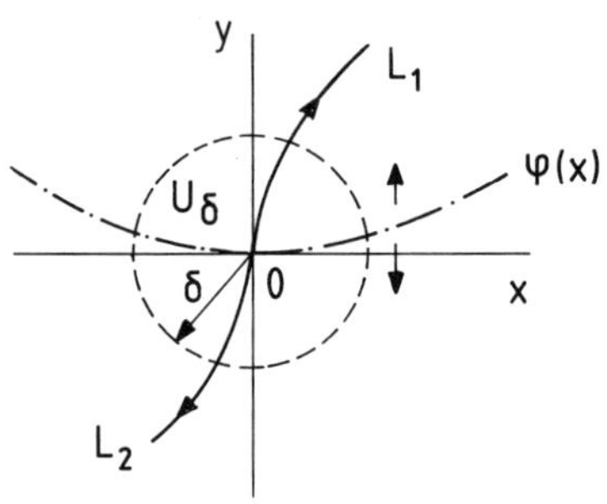

Fig. 1.8. Neighbourhood U_δ of O(0,0)

In the first case the positive and negative x-axes are obviously semipaths of the system, while in the second case ($\varphi(x) \neq 0$) the trajectories must cross the curve $y = \varphi(x)$. These trajectories are passed along in a direction depending on the sign of $\psi(x)$.

With the expression $\psi(x) = \Delta_m x^m + \ldots..$ we observe that this depends on the sign Δ_m and on the parity of m, which implies that four different cases have to be considered

($\Delta_m > 0$ or < 0, m is odd or even). According to Andronov, these considerations lead to a classification of topological structures which is summarized in Andronov's theorem 65 (p. 340, Andronov et. al, (1973)).

Theorem 65. Higher-order singular points with only one zero eigenvalue

Let O(0,0) be an isolated singular point of the system $\dot{x} = P_2(x,y)$ and $\dot{y} = y + Q_2(x,y)$ where P_2 and Q_2 are analytic and have series expansions near O(0,0) of which the lowest order terms are at least quadratic.
Let $y = \varphi(x)$ be the solution of $y + Q_2(x,y) = 0$ in the vicinity of O(0,0) and assume that the series expansion of the function $\psi(x) = P_2(x, \varphi(x))$ has the form $\psi(x) = \Delta_m x^m + \ldots$.. where $m \geq 2$ and $\Delta_m \neq 0$. Then:

1. If m is odd and $\Delta_m > 0$, O(0,0) is a topological node with an infinite number of semipaths in the directions O,π and exactly one semipath tending to 0 in the direction $\pi/2$ and also one in the direction $3\pi/2$.
2. If m is odd and $\Delta_m < 0$, O(0,0) is a topological saddle point whose separatrices (semipaths) approach O(0,0) in the direction $0,\pi/2$, π and $3\pi/2$, respectively.
3. If m is even, O(0,0) is a saddle-node, i.e. a singular point whose neighbourhood consists of one parabolic sector (nodal type) and two hyperbolic sectors (saddle-type). If $\Delta_m < 0$ the hyperbolic sectors contain a segment of the positive x-axis, if $\Delta_m > 0$ they contain a segment of the negative x-axis.

By definition the order of a singularity is equal to m.
The full proof of this theorem may be found in Andronov et. al (1973) pp. 337 ff. Typical phase plane pictures near these multiple-equilibrium points are shown in Fig. 1.9.

Fig. 1.9. Higher order singularities in $\mathbf{R}^2$ with one zero eigenvalue.

4.2 Higher-order singular points having both eigenvalues zero; p = 0

In this section we consider the system

$$\dot{x} = ax + by + P_2(x,y), \quad \dot{y} = cx + dy + Q_2(x,y)$$

with the assumption

$$p = a + d = 0$$

$$q = ad - bc = 0$$

$$|a| + |b| + |c| + |d| \neq 0$$

Suppose O(0,0) is an isolated singular point at the origin and $P_2(x,y)$ and $Q_2(x,y)$ are analytic in the vicinity of O. The series expansions of $P_2(x,y)$ and $Q_2(x,y)$ involve terms not lower than second order. Then a nonsingular linear transformation exists (c.f. Andronov p. 347) bringing the system into the normal form

$$\dot{x} = y + P_2(x,y) \qquad \dot{y} = Q_2(x,y)$$

The topological structure of O(0,0) is found by considering the isocline of vertical directions: $y = \varphi(x)$ where $\varphi(x)$ satisfies $\varphi(x) + P_2(x, \varphi(x)) = 0$.

On this curve the functions

$$\psi(x) = Q_2[x, \varphi(x)] \text{ and } p(x) = \frac{\partial P_2}{\partial x}[x, \varphi(x)] + \frac{\partial Q_2}{\partial y}[x, \varphi(x)]$$

may be evaluated and expanded as

$$\psi(x) = \Delta_m x^m + \ldots\ldots, \qquad p(x) = b_n x^n + \ldots\ldots$$

where Δ_m and b_n are the first nonvanishing coefficients in these expansions and m and n are integer.

The topological structure of phase portraits in $\mathbf{R}^2$ near singular points with zero as a double eigenvalue is then established by Andronov's theorems 66 and 67 (pp. 356, 362).

Theorem 66, 67. Higher-order singular points having both eigenvalues zero

Let O(0,0) be an isolated singular point of the system $\dot{x} = y + P_2(x,y)$, $\dot{y} = Q_2(x,y)$; let $y = \varphi(x)$ be the solution of $y + P_2(x, y) = 0$ near O(0,0) and assume that the series expansions of the function $\psi(x) = Q_2(x, \varphi(x))$ and $p(x) = \frac{\partial P_2}{\partial x}(x, \varphi(x)) + \frac{\partial Q_2}{\partial y}(x, \varphi(x))$ have the respective forms

$$\psi(x) = \Delta_m x^m + \ldots, \Delta_m \neq 0$$

$$p(x) = b_n x^n + \ldots, b_n \neq 0.$$

1. Let the number m be odd, $m = 2l + 1$ ($l \geq 1$) and $\lambda = b_n^2 + 4(l+1)\Delta_m$.
   Then if $\Delta_m > 0$ the singular point is a topological saddle-point, but if $\Delta_m < 0$ the following possibilities occur:

- a focus or a center if $b_n = 0$; if $b_n \neq 0$, $n > l$; if $b_n \neq 0$, $n = l$, $\lambda < 0$
- a topological node if $b_n \neq 0$, n is even, $n < l$; if $b_n \neq 0$, n is even, $n=l$, $\lambda \geq 0$
- a singular point with an elliptic region if $b_n \neq 0$, n = odd, $n < l$; if $b_n \neq 0$, n is odd, $n = l$, $\lambda \geq 0$.

2. Let the number m be even, $m = 2l$, $(l \geq 1)$.
   Then the singular point is
   - a saddle-node if $b_n \neq 0$, $n < l$
   - a cusp if $b_n = 0$; if $b_n \neq 0$, $n \geq l$.

The order of the singularity is by definition equal to m.
The proof of this theorem may be found in Andronov et. al (1973) pp. 365 ff.
Sketches of the phase portraits are shown in Fig. 1.10.

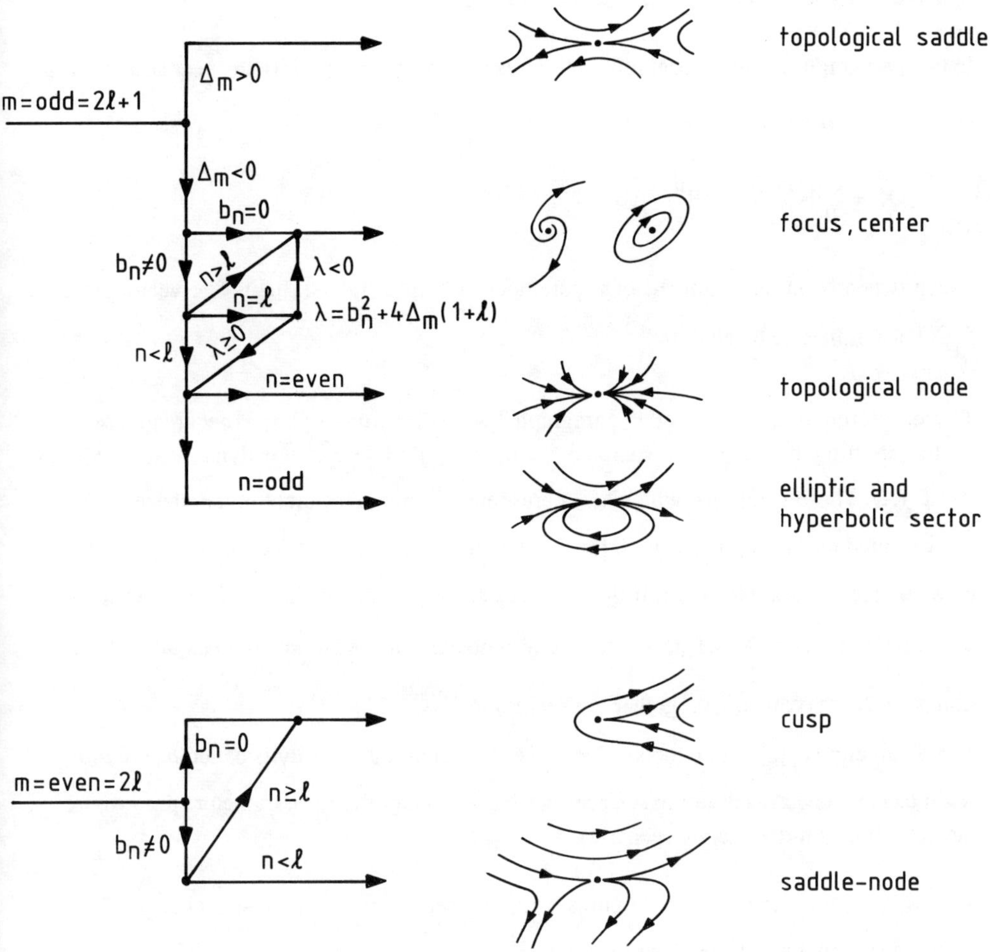

$$\dot{x} = y + P_2(x, y),\ \dot{y} = Q_2(x, y)$$

$$\dot{y}\Big|_{\dot{x}=0} = \Psi(x) = \Delta_m x^m \ldots\ldots.$$

$$\frac{\partial P_2}{\partial x} + \frac{\partial Q_2}{\partial y}\Big|_{\dot{x}=0} = p(x) = b_n x^n + \ldots\ldots$$

Fig. 1.10. Higher-order singularities in $\mathbf{R}^2$ with two eigenvalues zero.

5. Bifurcation of vector fields, unfoldings

In this paragraph we study local bifurcations as they occur in vector fields near structurally unstable singular points. Consider a vector field in $\mathbf{R}^n$ defined as

$$\dot{X} = f_\mu(X), \; X \in \mathbf{R}^n, \;\; \mu \in \mathbf{R}^k$$

which depends on the k-dimensional parameter $\mu = (\mu_1, \mu_2, \ldots, \mu_k)$. The vector field $f_\mu(X)$ is assumed to be analytic.

The term bifurcation introduced in paragraph 3 was originally used by Poincaré to describe the splitting behaviour of stationary solutions: $X = X_o$ of the dynamical system $X = f_\mu(X)$. These solutions, which are represented as singular points in the phase space, can be found by solving $f_\mu(X_o) = 0$ for X_o as a function of μ if the Jacobian $Df_\mu(X_o)$ has no zero eigenvalues. However, if the Jacobian has a *zero* eigenvalue, occurring at some μ, say μ_c, the system $\dot{X} = f_\mu(X)$ is structurally unstable and several branches of the solution $X_o = X_o(\mu)$ can come together at (X_o, μ_c) in $\mathbf{R}^{n+k}$.
The parameter set μ_c where the system looses its structural stability is called *bifurcation set*, it can be viewed as dividing surfaces bordering domains in the parameter space where the system is non-degenerate (generic).

Variations in the parameter space intersecting the set $\{\mu \mid \mu = \mu_c\}$ cause a change of the topological structure of the phase portrait.
Such changes are called bifurcations and the corresponding parameter values are the bifurcation values.
A *one*-parameter family of systems with k-1 relations between the parameters μ_1, $\mu_2, \ldots, \mu_k$ is represented at a curve λ in the k-dimensional parameter space. Assume that λ intersects the bifurcation set, μ_c, and that the intersection is *transversal*. At the intersection, structural unstability occurs, but due to the transversality condition the degeneracy can be removed by any small displacement along λ. The corresponding perturbation is called a *generic* perturbation. Variations in the parameter space that coincide with μ_c cor-

respond to a *nongeneric* perturbation.

In this paragraph we focus the attention on bifurcations of isolated singular points. As we have seen these bifurcations are referred to as local bifurcations, so that the vector field is studied near degenerate singular points and the bifurcating solutions are also found in the neighbourhood of these points. Let us start with a simple example of a dynamical system in $\mathbf{R}^1$. Consider the one-dimensional 'vector' field:

$$\dot{x} = f_\mu(x) = \mu x - x^3 \tag{1.8}$$

where μ is now a scalar parameter.

Here $Df_\mu(x) = \mu - 3x^2$, and the only bifurcation point is $(x_o, \mu_o) = (0,0)$. It is easy to check that $f_\mu(x) = 0$ has one solution if $\mu \leq 0$ ($x_o = 0$) and three different solutions if $\mu > 0$ ($x_o = 0, \pm\sqrt{\mu}$).

A qualitative picture of these solutions is given in Fig. 1.11a which shows the branches of singular points in a (x, μ) space. This figure is called the bifurcation diagram; it shows the loci of singular points as a function of the parameters.

The phase space is one-dimensional and coincides with the x-axis.

The point $x = 0$ is stable for $\mu \leq 0$ and it becomes unstable for $\mu > 0$.

The phase portraits which can occur after bifurcation are shown in Fig. 1.11b.

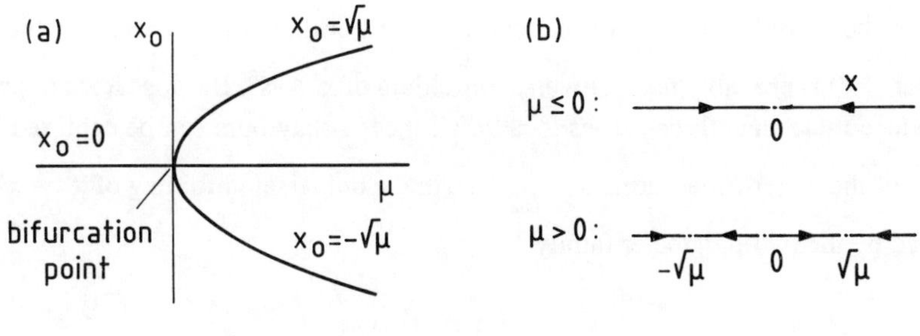

Fig. 1.11. Bifurcation of $f_\mu(x) = \mu x - x^3$.

(a) bifurcation diagram, (b) phase portraits.

The previous example of a one-parameter bifurcation gives the result of a one-parameter variation on the structurally unstable 'vector' field $\dot{x} = -x^3$.

The perturbed system $\dot{x} = +\mu x - x^3$ shows some possible bifurcations and is called an *unfolding* of the unperturbed system $\dot{x} = -x^3$. Now the important question arises: is the unfolding given by eq. (1.8) a particular one, showing only some bifurcations, or has the unfolding a more general character so that all possible bifurcations of $\dot{x} = -x^3$, within a certain class of perturbations (for example analytic), are embodied. If the unfolding describes all possible bifurcations of the degenerate singular point, then it is called a universal unfolding.

Hence a universal unfolding of a degenerate vector field is a family of bifurcating solutions which contains the bifurcation in a persistent way. It has the important property that it describes the bifurcation with the smallest number of parameters. This number is called the *codimension* of the degenerate vector field. The codimension of a bifurcation is the smallest dimension of the parameter space which contains the bifurcation in all its aspects. A universal unfolding is not necessarily unique since a coordinate transformation can lead to a differently formulated unfolding having the same dimension. Therefore the term 'universal unfolding' is not generally accepted in the literature; for example Shirer and Wells (1983) prefer the term *versal unfolding* to point out the nonuniqueness of the unfolding.

Bearing this in mind we return to the system $\dot{x} = -x^3$ in order to find the universal unfolding of this system. Eq. (1.8) gives an unfolding with one parameter, but does it describe the bifurcation in all its aspects or are more parameters necessary for a universal unfolding?

If the function $f(x) = -x^3$ is subjected to a small perturbation $f(x) \rightarrow f_\mu(x)$, it is obvious that $f_\mu(x)$ possesses one, two or three zeroes near $x = 0$.

However, the unfolding (1.8) gives either one ($\mu < 0$) or three ($\mu > 0$) zeroes, which indicates that (1.8) is probably not a universal unfolding of $\dot{x} = -x^3$. Because it is not possible to introduce more than three zeroes locally, all these behaviours can be captured by the addition of the lower-order terms $\mu_1 + \mu_2 x$, so that a universal unfolding of $\dot{x} = -x^3$ is represented by the two-parameter family

$$f_\mu(x) = \mu_1 + \mu_2 x - x^3 \quad *) \tag{1.9}$$

Equating $f_\mu(x)$ to zero gives the loci of singular points x_o as a function of the parameters μ_1 and μ_2. Figure 1.12 shows the corresponding bifurcation diagram with these loci in the x_o, μ_1, μ_2 space.

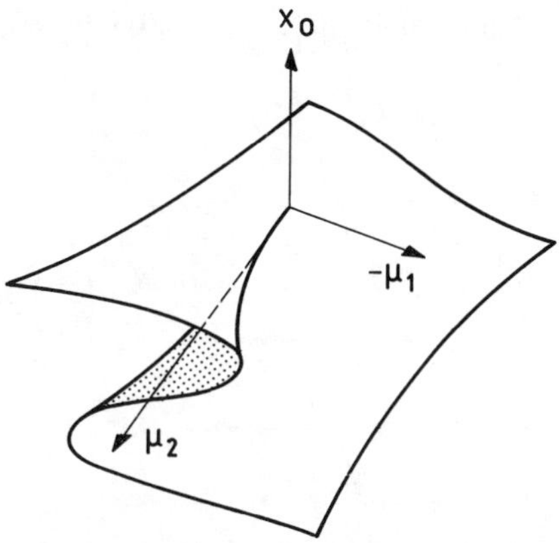

Fig. 1.12. Bifurcation diagram of $f_\mu(x) = \mu_1 + \mu_2 x - x^3$.

A singular point becomes degenerate if $Df(x_o)$ has eigenvalues with zero real part, thus if $\mu_2 - 3x_o^{\,2} = 0$.

Eliminating x_o from this equation and the equation $f_\mu(x_o) = 0$ we find the parameter combinations at which bifurcation takes place.

These parameter combinations are called the *bifurcation set* which satisfies:

$$(\mu_2/3)^3 - (\mu_1/2)^2 = 0$$

---

*) Other universal unfoldings such as $f_\mu(x) = \mu_1 x + \mu_2 x^2 - x^3$ or $f_\mu(x) = \mu_1 + \mu_2 x^3 - x^3$ have the same bifurcation characteristics; they can be transformed into eq. (1.9) by a suitable translation of the x-coordinate.

This set consists of two branches if $\mu_2 > 0$, the corresponding one-dimensional phase portraits are shown in Figure 1.13. We observe from the phase portraits that on a branch ($\mu_2 > 0$) the nonhyperbolic point is the one-dimensional variant of a saddle-node. Crossing a branch transversally implies the bifurcation of the saddle-node. Such a crossing is in fact described by one parameter so that the branches represent a *codimension-one* bifurcation. At the point $(\mu_1, \mu_2) = (0, 0)$ where both branches terminate with a common tangent we have a nonhyperbolic one-dimensional node; its bifurcation is described by two parameters implying that the point $(\mu_1,\mu_2) = (0,0)$ represents a *codimension-two* bifurcation.

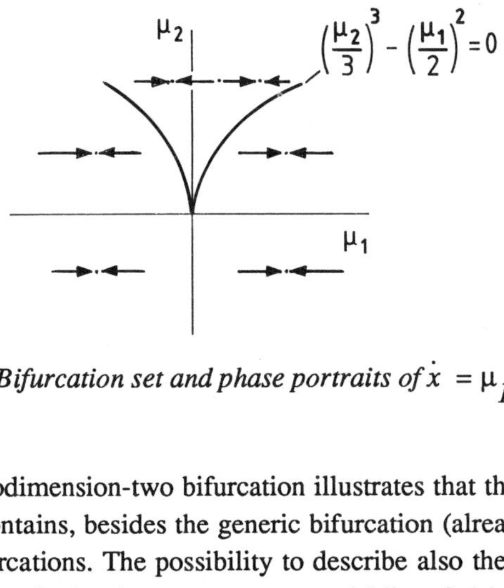

Fig.1.13. Bifurcation set and phase portraits of $\dot{x} = \mu_1 + \mu_2 x - x^3$.

This example of a codimension-two bifurcation illustrates that the unfolding of $\dot{x} = -x^3$ given by eq. (1.9) contains, besides the generic bifurcation (already given by eq. (1.8)), also nongeneric bifurcations. The possibility to describe also these nongeneric bifurcations is realized by introducing the two parameter unfolding: $f_\mu(x) = \mu_1 + \mu_2 x - x^3$, implying that the universal unfolding of $\dot{x} = -x^3$ is defined with two parameters involving a codimension-two bifurcation. If only generic bifurcations are of interest it is enough to consider the one-parameter unfolding as given by eq. (1.8).

The preceding consideration about bifurcations of a one-dimensional system illustrates that concepts like unfolding and codimension can be valuable tools in bifurcation analyses. Especially if the aim is to develop a general theory about degeneracies and their uni-

versal unfoldings, the concept codimension plays an important role. It offers the possibility to develop a strategy for the study of these degeneracies whereby one starts by investigating codimension-one bifurcation, then codimension-two bifurcations etc. Such a strategy can ultimately lead to a classification scheme of bifurcations based on general considerations such as number of parameters, dimension of the phase space and constraints which account for the class of perturbations, see for example Shirer and Wells (1983).
In the next table we have summarized a few examples of elementary universal unfoldings of degenerate singularities occurring in one- and two-dimensional systems.

One-dimensional universal unfoldings

form	codimension	unfolding	name
$-x^2$	1	$\mu_1 - x^2$	fold
$-x^3$	2	$\mu_1 + \mu_2 x - x^3$	cusp
$-x^4$	3	$\mu_1 + \mu_2 x + \mu_3 x^2 - x^4$	swallowtail
$-x^5$	4	$\mu_1 + \mu_2 x + \mu_3 x^2 + \mu_4 x^3 - x^5$	butterfly

Two-dimensional universal unfoldings

form	codimension	unfolding	name
$\begin{pmatrix} xy \\ x^2 - y^2 \end{pmatrix}$	4	$\begin{pmatrix} \mu_1 + \mu_4 y + xy \\ \mu_2 + \mu_3 y + x^2 - y^2 \end{pmatrix}$	hyperbolic-umbilic
$\begin{pmatrix} x^2 \\ y^2 \end{pmatrix}$	4	$\begin{pmatrix} \mu_1 + \mu_3 x + \mu_4 y + x^2 \\ \mu_2 + y^2 \end{pmatrix}$	elliptic-umbilic

As we have mentioned before, the presented form of the unfoldings is not necessarily unique; several alterations can be found which account for the same bifurcation behaviour. These alterations follow from suitable coordinate transformations and rearrangement of the parameters.
The loci of the nearby singularities: $x_o(\mu_1, \mu_2, \ldots, \mu_k)$, forming the bifurcation diagram can be portrayed as surfaces (boundaries) in the x_o - μ-space ($\mu \in \mathbf{R}^k$). The complex geometrical structure of these surfaces near the origin can be associated with well known elementary singularities appearing in catastrophe theory. This correspondence with ele-

mentary catastrophes is reflected by the names listed in the table.

6. Center manifolds

Before proceeding with the applications in fluid flow problems, there is a general technique which has to be discussed first, as it can simplify the analysis of bifurcation problems considerably. This technique has the effect of introducing coordinate systems in which computations are more easily carried out. After using it one is left with a reduced system of differential equations containing all the qualitative features of the bifurcation. It must be emphasized that this technique, called *center manifold theory*, has a local character and is only applicable to bifurcations of singular points.

The center manifold theory provides a means for systematically reducing the dimension of the state spaces which need to be considered when analyzing bifurcations of a given type.

Suppose we have a system of nonlinear ordinary differential equations:

$$\dot{X} = f(X) \qquad X \in \mathbf{R}^n$$

for which $f(0) = 0$.

The linearized system at $X = 0$ may be written as

$$\dot{X} = Df(0).X$$

If $Df(0)$ has no eigenvalues with zero real part, then Hartman-Grobman's theorem states that the eigenvalues, with positive and negative real parts, determine the local phase portrait near $X = 0$. If there are eigenvalues with zero real parts the topological structure can be quite complicated. Some examples in $\mathbf{R}^2$ have already been given in this chapter (paragraph 4).

A solution of the linearized system, satisfying the initial condition $X(t_o)=X_o$, $X_o \in R^n$ is the vector valued function

$$X(t; X_o) = \exp(t\, Df(0))\, X_o$$

A general solution of $\dot{X} = Df(0)\,X$ is obtained by linear superposition of n linearly independent solutions: $\{V^1(t), \ldots V^n(t)\}$

$$X(t) = \sum_{j=1}^{n} C_j\, V^j(t)$$

where C_j $(j = 1, \ldots n)$ is a constant.

A set of linearly independent solutions can be obtained from the real and imaginary parts of the vector valued functions $\exp(\lambda_j t)\, X^j$ where X^j are the (generalized) eigenvectors associated with the eigenvalues (real or complex) $\lambda_j (j = 1, \ldots n)$.
Various subspaces spanned by (generalized) eigenvectors and filled with solution curves may be distinguished.

the stable subspace, E^s: spanned by $X_s^1 \ldots X_s^{n_s}$

the unstable subspace, E^u: spanned by $X_u^1 \ldots X_u^{n_u}$

the center subspace, E^c: spanned by $X_c^1 \ldots X_c^{n_c}$

where:

$X_s^1, \ldots X_s^{n_s}$ are n_s (generalized) eigenvectors with eigenvalues having *negative* real parts

$X_u^1, \ldots X_u^{n_u}$ are n_u (generalized) eigenvectors with eigenvalues having *positive* real parts

$X_c^1, \ldots X_c^{n_c}$ are n_c (generalized) eigenvectors with eigenvalues having *zero* real parts

Since the dimension of the whole phase space is n we have $n_s + n_u + n_c = n$.
The adjectives stable, unstable and center reflect the behaviour of solutions in their eigenspaces respectively.

Solutions lying in E^s are characterized by exponential decay (monotonic or oscillatory), those lying in E^u by exponential growth and those lying in E^c by neither. In Fig. 1.14 we show some examples of invariant subspaces in $\mathbf{R}^2$ and $\mathbf{R}^3$.

For nonlinear systems one can define the invariant subspaces W^s and W^u being the nonlinear analogues of E^s and E^u respectively. The subspaces W^s and W^u are frequently denoted as stable and unstable manifolds, they are tangent to E^s and E^u at the singular point where $f(X) = 0$.
The existence of these manifolds for nonlinear systems is established by the stable manifold theorem.

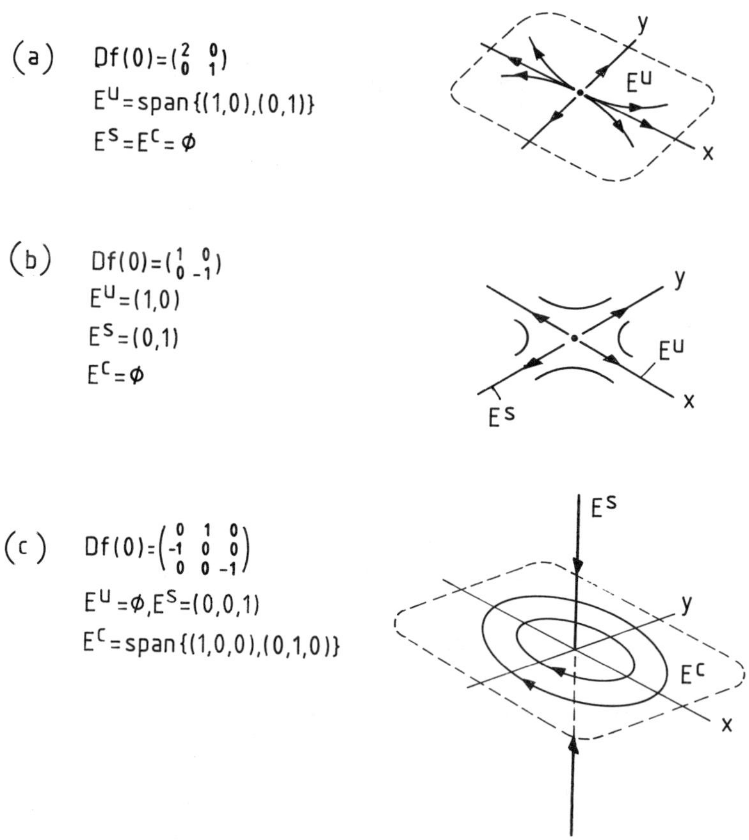

Fig. 1.14. Subspaces of linear flows.

Stable Manifold Theorem for hyperbolic singular points.

Suppose that $\dot{X} = f(X)$ has a hyperbolic singular point X_o, then there exist stable and unstable manifolds $W^s(X_o)$, $W^u(X_o)$, of the same dimensions n_s, n_u as those of the eigenspaces E^s, E^u of the linearized system; at X_o, $W^s(X_o)$ is tangent to E^s and $W^u(X_o)$ is tangent to E^u; $W^s(X_o)$ and $W^u(X_o)$ are as smooth as the function f.

The proof of this theorem has been given by Hartman (1964) and more recently by Carr (1981).

We note that this theorem says nothing about the existence of center manifolds W^c for nonlinear systems. For the case that the nonlinear system is C^r the existence of center manifolds is established in the center manifold theorem which includes the results of the stable manifold theorem. The first proof of the center manifold theorem has been given by Kelley (1967).

Center Manifold Theorem

Let f be a C^r vector field on $\mathbf{R}^n$ vanishing at $X = X_o$ so that $f(X_o) = 0$.

The spectrum of eigenvalues $\lambda_1, \dots, \lambda_n$ of $Df(X_o)$ is divided into three parts, $\sigma_s, \sigma_u, \sigma_c$ with

$$\lambda \in \sigma_s \quad \text{if} \quad \operatorname{Re}\lambda < 0$$

$$\lambda \in \sigma_c \quad \text{if} \quad \operatorname{Re}\lambda = 0$$

$$\lambda \in \sigma_u \quad \text{if} \quad \operatorname{Re}\lambda > 0$$

The corresponding generalized eigenspaces are E^s, E^c and E^u respectively. Then there exist C^r stable and unstable manifolds W^s and W^u tangent to E^s and E^u at X_o; respectively and a C^{r-1} center manifold W^c tangent to E^c at X_o. The manifolds W^s, W^u and W^c are each filled with solution curves.

The center manifold theorem implies that a nonlinear system f(X) can be split up, near a singular point X_o, into two subsystems f^{s+u} and f^c of which the linear parts have eigenvalues with nonzero and with zero real parts, respectively.

Then one may write the system $\dot{X} = f(X)$ near the singular point X_o: (X_o,Y_o) formally as

$$\dot{X} = f^c(X,Y) = Df^c(X_o,Y_o) \,.\, (X-X_o) + F(X,Y)$$
$$\dot{Y} = f^{s+u}(X,Y) = Df^{s+u}(X_o,Y_o) \,.\, (Y-Y_o) + G(X,Y) \tag{1.10}$$

where $X \in \mathbf{R}^{n_c}$, $Y \in \mathbf{R}^{n_s+n_u}$ and F, G are nonlinear higher order terms which vanish at (X_o,Y_o).

Since the center manifold is tangent to E^c at (X_o,Y_o) it can be written explicitely

$$Y\text{-}Y_o = h(X) \quad \text{with } h(X_o) = Dh(X_o) = 0$$

The center manifold W^c may be viewed as a particular set of solutions of the original system $\dot{X} = f(X)$, satisfying the additional constraints $f(X_o) = Df(X_o) = 0$.

They may be calculated by substituting $Y = Y_o + h(X)$ into eq. (1.10), resulting in

$$D\,h(X)\,[Df^c(X_o,Y_o)(X-X_o) + F(X,Y_o+h(X))]$$
$$-Df^{s+u}(X_o,Y_o)h(X) - G(X,Y_o+h(X)) = 0 \tag{1.11}$$

with the boundary conditions $h(X_o) = Dh(X_o) = 0$.

Equation (1.11) is in general a partial differential equation which cannot be solved exactly in most cases, but its solution can sometimes be approximated by a series expansion near X_o. If such an approximate solution is found, the original system may be projected on the center manifold resulting into the following reduced system

$$\dot{X} = Df^c(X_o,Y_o) \,.\, (X-X_o) + F(X,Y_o + -h(X)) \tag{1.12}$$

Since h(X) is tangent to E^c at the singular point, the solutions of the reduced system provide an approximation of (1.10) near X_o. This conclusion was obtained, independently, by Henry (1981) and Carr (1981) when they proved that the properties of (1.10) near (X_o,Y_o) and of (1.12) near X_o are the same.

It implies that the bifurcation properties of a degenerate singularity can be obtained by analysing the nonlinear system on its center manifold. Although this method looks straightforward, its succes depends largely on the possibility whether a center manifold can be found or at least approximated in a suitable way.

Other difficulties that can appear are the non-uniqueness and loss of smoothness of solutions in the center manifold. The boundary conditions $h(X_o) = Dh(X_o) = 0$ are not sufficient for a unique solution of eq. (1.11).

In the following examples we point out some of these aspects (possibilities, limitations) of the use of center manifolds.

In the first example center manifold theory will be used in order to find the phase portraits of the two-dimensional system

$$\dot{x} = \varepsilon x + xy - x^3$$

$$\dot{y} = -y + y^2 - x^2$$

if the real-valued parameter ε varies near zero.

If ε = 0 the system has a degenerate singular point at the origin (0,0); the corresponding coefficient matrix of the linear part is given by $\begin{bmatrix} 0 & 0 \\ 0 & -1 \end{bmatrix}$. The eigenspaces are: $E^s = (0,1)$, $E^c = (1,0)$, $E^u = \phi$.

Using theorem 65 (paragraph 4) the singularity at the origin is a stable topological node; an infinite number of paths approach the node along the x-direction.

To obtain the bifurcation solutions for ε ≠ o we consider the center manifold of the suspended system in $\mathbf{R}^3$

$$\begin{aligned} \dot{x} &= \varepsilon x + xy - x^3 \\ \dot{\varepsilon} &= 0 \\ \dot{y} &= -y + y^2 - x^2 \end{aligned} \tag{1.13}$$

where the 'parameter' ε is seen as 'trivial' dependent variable.

The coefficient matrix of the linear part becomes for $\varepsilon = 0$

$$\begin{pmatrix} 0 & 0 & 0 \\ 0 & 0 & 0 \\ 0 & 0 & -1 \end{pmatrix} \quad \text{with eigenvalues } \lambda_x = \lambda_\varepsilon = 0,\ \lambda_y = -1$$

The corresponding eigenspaces are $E^c = (1,0,0) \times (0,1,0)$, $E^s = (0,0,1)$ and $E^u = \phi$. Now we seek a center manifold $h = h(x,\varepsilon)$ tangent to E^c, which is a solution of eq. (1.11), satisfying the boundary conditions $h(0,0) = \frac{\partial h}{\partial x}(0,0) = \frac{\partial h}{\partial \varepsilon}(0,0) = 0$.
Equation (1.11) becomes in this case

$$\left[\frac{\partial h}{\partial x}, \frac{\partial h}{\partial \varepsilon}\right]\left[\begin{matrix} \varepsilon x - x^3 + xh \\ 0 \end{matrix}\right] + h - h^2 + x^2 = 0$$

The center manifold will be approximated by

$$h(x,\varepsilon) = \alpha_1 x^2 + \alpha_2 x\varepsilon + \alpha_3 \varepsilon^2 + \beta_1 x^3 + \beta_2 x^2 \varepsilon + \beta_3 x\varepsilon^2 + \beta_4 \varepsilon^3 + O(4)$$

where $O(4)$ denotes terms of $O[(|x| + |\varepsilon|)^4]$.

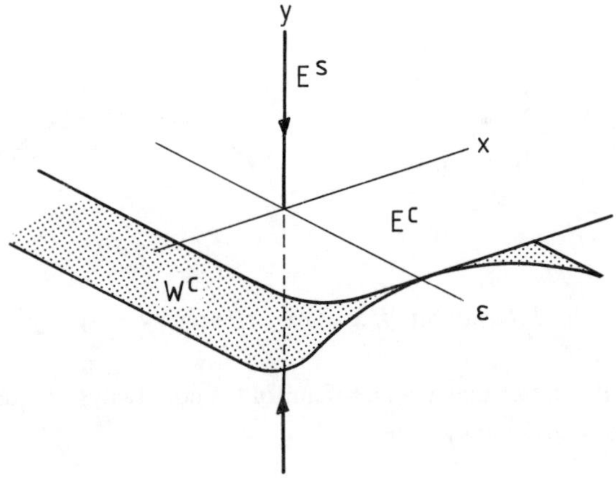

Fig. 1.15. Center manifold W^c*:* $h(x,\varepsilon) = -x^2 + 2x^2\varepsilon + O(4)$.

Equating powers of x^2, $x\varepsilon$, ... ε^3 we find for the center manifold (see Fig. 1.15)

$$W^c: h(x,\varepsilon) = -x^2 + 2x^2\varepsilon + O(4)$$

The reduced system which governs the bifurcation caused by ε, follows as the projection of system (1.13) on W^c:

$$\dot{x} = \varepsilon x - 2x^3 + 2\varepsilon x^3 + O(4), \ \dot{\varepsilon} = 0$$

The vector field on W^c is a one-parameter unfolding of the one-dimensional degeneracy: $-2x^3$.

The unfolding has also a singularity at $x = 0$ and two neighbouring singular points at $x = \pm\sqrt{\varepsilon/2} + O(\varepsilon^{\frac{3}{2}})$. The singular point $x = 0$ appears to be stable for $\varepsilon < 0$ and unstable for $\varepsilon > 0$. The behaviour of the reduced system on W^c enables us to draw the trajectories in the x,y-phase plane if ε varies near zero. Since the unstable eigenspace E^u appeared to be empty, the trajectories tend to W^c for increasing t. see Fig. 1.16.

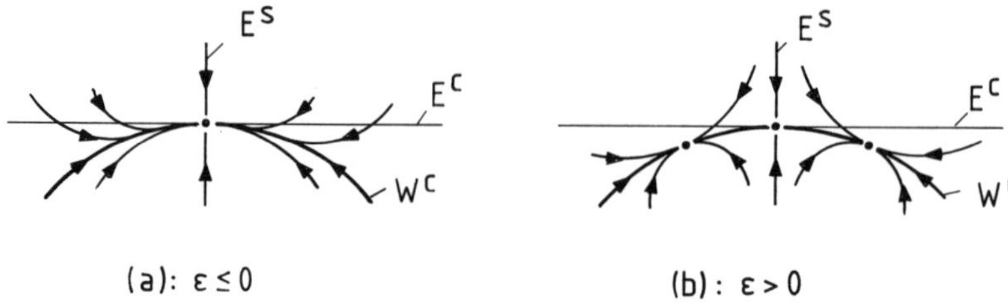

Fig. 1.16. Bifurcation of: $\dot{x} = \varepsilon x - x^3 + xy,\ \dot{y} = -y + y^2 - x^2.$

The next example illustrates that a center manifold is not always unique.
Consider the two-dimensional system

$$\dot{x} = -x^3, \qquad \dot{y} = -y$$

which has a isolated singular point (topological node) at the origin. The center eigenspace E^c is formed by (1,0) and the center manifold W^c: h(x) is a solution of

$$-x^3 h' + h = 0 \quad \text{with } h(o) = h'(o) = 0$$

The general solution of this differential equation is

$$h(x) = C.\exp(-\frac{1}{2x^2})$$

Since the solution satisfies h(o) = h'(o) = 0 irrespective the value of C, it must be concluded that an infinite number of center manifolds exists.

Since all derivatives of $\exp(-\frac{1}{2x^2})$ vanish at $x = 0$ only the center manifold $h = 0$ is analytic, and all other center manifolds are C^∞.

7. An approach to physical unfoldings in flow patterns

As announced in the introduction, this work deals with bifurcations in vector fields as they arise as velocity fields in gas- and fluid motions; e.g. conical flows (Chapter II) and viscous flows retarded near a smooth surface (Chapter III). In general at stagnation points these vector fields will have isolated singularities which can degenerate under certain conditions. To obtain a universal unfolding of such a degeneracy one can follow a mathematical and a physical approach. The former, being very straightforward, determines the *universal unfolding* of a degenerate state on the basis of its topological features such as codimension, center manifold and on the basis of the class of perturbations admitted.
In this way, the complete bifurcation can be described with the smallest number of parameters (say μ_1, ... μ_k, $k \in \mathbf{R}$) and general unfolding principles can be used.

Although this approach is very attractive its application involves problems of different kind.

First, degeneracies occurring in flow fields appear scarcely in those convenient forms being necessary for a straightforward application of unfolding theory. These degeneracies - together with its unfoldings - are more or less of particular type because the corresponding degenerate solutions have to satisfy the partial differential equations which govern the physics of the flow (e.g. conservation of mass, momentum and energy).
Consequently, it means that we are less interested in finding the universal unfolding of the degeneracy but more in those unfoldings which are allowed by the flow equations. Let us call such a permitted unfolding a *physical* unfolding.

A second difficulty stems from the problem of interpreting the involved parameters of the universal unfolding.
The physical interpretation of these parameters is a necessary but by no means easy task, as the physically relevant parameters may be related to the bifurcation parameters μ_1, ... μ_k in a very complicated way. Furthermore, the number of relevant physical parameters may exceed the number of bifurcation parameters so that a single bifurcation parameter may be constituted by a combination of several physical parameters. To find these combinations, one needs additional information from the flow problem which cannot be derived from the universal unfolding.
On the other hand, if the number of relevant physical parameters is smaller than the codimension, it implies the existence of a dependency between the bifurcation parameters. This dependency can significantly reduce the possible bifurcation solutions with respect to those given by the universal unfolding. Moreover if the number of physical parameters

is not known beforehand there is a risk that unallowed bifurcations will creep in, which remain unnoticed, and lead to a distorted view about transition behaviour in flow patterns.

In a physical approach, the above mentioned difficulties are circumvented since a degenerate state in the flow pattern and the corresponding physical unfolding appear simultaneously if the governing flow equations are evaluated near singular points. Actually, the evaluation results into approximate solutions which satisfy these equations to a certain order near the singularity. These approximations contain several unknowns which cannot be determined in a local analysis.
For specified values of the unknowns the singularity attains a degenerate state indicating that the involved unknowns can be viewed as bifurcation parameters. Then the physical unfolding of the degenerate singularity follows quite easily by varying the involved parameters with respect to bifurcation values at which the degenerate singularity was found.

The author is inclined to favour the physical approach as long as the emphasis lies on local bifurcations in vector fields that are generated by physical systems.

8. References

Andronov, A.A., Leontovich, E.A., Gordon, I.I. and Maier, A.G. (1973)
Qualitative theory of second-order dynamical systems. Wiley, New York.

Coddington, E.A. and Levinson, L. (1955)
Theory of ordinary differential equations. McGraw-Hill, New York.

Carr, J. (1981)
Applications of center manifold theory. Springer-Verlag, New York, Berlin.

Guckenheimer, J. and Holmes, P.J. (1983)
Nonlinear oscillations, dynamical systems and bifurcation of vector fields. Springer-Verlag, New York, Berlin, Heidelberg, Tokyo.

Hartman, P. (1964)
Ordinary differential equations. Wiley, New York.

Henry, D. (1981)
Geometric theory of semilinear parabolic equations. Lecture notes in mathematics, Vol. 840. Springer-Verlag, New York, Heidelberg, Berlin.

Jordan, D.W. and Smith, P. (1977)
Nonlinear ordinary differential equations. Clarendon Press, Oxford.

Kelley, A. (1976)
The stable, center stable, center, center unstable and unstable manifolds. J.Diff. Eqns, 3, 546-570.

Reyn, J.W. (1964)
Classification and description of the singular points of a system of three linear differential equations. J. Appl. Math. Phys., 15, 540-557.

Shirer, H.N. and Wells, R. (1983)
Mathematical structure of the singularities at the transitions between steady states in hydrodynamic systems. Lecture notes in physics, Vol. 185. Springer-Verlag, Berlin, Heidelberg, New York, Tokyo.

CHAPTER II. TOPOLOGY OF CONICAL FLOW PATTERNS

1. Introduction

1.1. Concepts and definitions

The qualitative theory of dynamical systems will be applied to three-dimensional inviscid flows with conical symmetry. Such flows, which are called conical flows, have the specific property that the velocity and the quantities defining the state of the gas, e.g. pressure and temperature, are constant along rays emanating from a common point in the physical space. This point is called the center of the conical field.
Inviscid conical gas flows embody an important class of flows around conical configurations of practical interest, such as delta wings having arbitrary cross-section, inlet configurations, nose cones, wing-body junctions and internal flows in nozzles and diffusors.

Due to the conical symmetry, these flows can be described by functions of two independent variables defining the position of the rays. The structure of a conical flow may be represented on a unit sphere centered at the conical center. The velocity vector $\underline{V}$ may be decomposed into a radial component $\underline{V}_r$ normal to the unit sphere and a component $\underline{V}_t$ tangent to it; the latter defines a vector field on this sphere. Integration of this vector field yields lines on the unit sphere which will be called *conical streamlines*. The physical significance of a conical streamline may be explained quite easily. Consider the spatial streamlines passing through a common ray. Since the flow is assumed to be conical these streamlines form a conical streamsurface, which is built up by sequential rays emanating from the conical center. Taking the intersection of a conical stream surface with the unit sphere, a conical streamline, is obtained.
Similarly, conical streamlines can be defined on a plane surface Σ at unit distance from the conical center. Such a plane is usually called a cross flow plane and the conical streamlines follow as the intersections of conical streamsurfaces with Σ. Analogous to the situation on the unit sphere, conical streamlines in the cross flow plane Σ can be taken as integral curves of a vector field in Σ. This vector field may be obtained if the velocity vector $\underline{V}$ is decomposed nonorthogonally into a radial component and into a component in the cross flow plane. The latter, which is called cross flow velocity, yields the vector field in the cross flow plane.

Points in the cross flow plane and on the unit sphere where the cross flow velocity vanishes, thus the velocity has only a radial component, are called *conical stagnation points*. These points in the vector field are critical points where the conical streamline direction is undetermined.

1.2. A brief survey of conical flow theory

The early theoretical work on conical flows is largely based on linearized theory and originates from the work of Busemann (1929) on circular cones in supersonic flow. From that time *linear* theory of conical flows is developed and applied to configurations of aerodynamic interest by several authors. Much of the early work culminated in the studies of Germain (1949) and Goldstein & Ward (1950). The development of a nonlinear theory for conical flows starts with the work of Taylor & Maccoll (1933) when they studied the axisymmetric flow around a circular cone.

A great variety of theoretical aspects of nonlinear conical flows is treated by Bulakh in the late fifties. He concentrated his efforts on the formulation of boundary value problems and on finding approximate solutions for them. These solutions are discussed by Bulakh with special emphasis to their properties and singular behaviour, in particular on the Mach-cone and near shock waves. A further study on the properties of nonlinear isentropic conical flows is performed by Reyn (1960), where he studied the solution of the nonlinear partial differential equation from the point of view of differential geometry. In doing so Reyn commented and clarified, certain discrepancies, occurring in the flow around a flat delta wing with supersonic leading edges as given by Maslen (1952) and Fowell (1956).

All these investigations contributed significantly to the theory of nonlinear conical flows which, to a certain extent, has reached a final form by the publication (in Russian) in 1970 of Bulakh's book: 'Nonlinear Conical Gas Flows'.

Ever since that time the main attention diverts to the application of numerical methods to obtain solutions for flows around cones and delta wings at incidence.

Although these methods give accurate solutions at moderate incidence, for larger angles of attack two fundamental difficulties arise which prevent a straightforward numerical solution.

First, as the angle of attack is increased, the cross flow velocity will increase to supersonic values in some parts of the flow field. This changes the nature of the governing partial differential equations from elliptic to a mixed elliptic-hyperbolic type and results in the appearence of a conical supersonic region terminated by an embedded shock wave.

Another difficulty is that some flow properties such as entropy, density and radial velocity can become multivalued at conical stagnation points. Since entropy gradients can be identified with rotationality of the flow (Crocco's theorem), these conical stagnation points can appear as vortical singularities. The presence of vortical singularities, which was a severe obstacle for a straightforward numerical calculation, see for example Stocker and Mauger (1962), did revive the attention to nonlinear conical flow theory with special emphasis on the conical flow structure near conical stagnation points. It is of value therefore, to evaluate as systematically as possible by means of a local analysis, the possible flow structures near such points, so that in a particular flow problem the qualitative basis for a numerical procedure can be selected with more certainty.

In conical flows with entropy gradients the entropy remains constant on conical streamlines. Then, if in a conical stagnation point various streamlines merge, a vortical singularity is formed. This conjecture was put forward by Ferri (1951) when he discussed the supersonic flow past a circular cone at incidence.

Since the appearance of Ferri's paper investigations of the flow near conical stagnation points show an emphasis of interest in the possible conical streamline patterns together with the related pressure distributions near such points. Melnik (1967) constructed some approximate solutions of the nonlinear inviscid conical flow equations near conical stagnation points attached to a body surface. These solutions involve entropy gradients in the flow. When the streamline pattern is related to the corresponding pressure distribution on the body surface no unique correspondence was found. Bakker (1977) showed that for these solutions a unique correspondence may be obtained if the pressure distribution normal to the body surface is also taken into account. Both investigations indicate that the presence of entropy gradients does not affect the topological structure of the conical streamline pattern that corresponds to a given pressure distribution. This result was further confirmed in the special case of the conical stagnation points in the flow past slender circular cones at high incidence, as calculated using slender body theory (Smith 1972), or linearized theory (Bakker & Bannink 1974).

It is of interest, therefore, to make a further study of the topological structure of conical flows near conical stagnation points using the assumption of potential flow. An advantage of this approach is that the nonlinear conical flow equations reduce to a single second-order equation for the conical potential for which solutions are simpler to obtain. Moreover in a conical stagnation point. this equation becomes Laplace's equation which is also satisfied by the velocity potential in incompressible plane flow. Stagnation-point solutions for incompressible plane flows are then used as a guide to conical stagnation point solutions. Since conical stagnation points appear as critical points in a vector field in $\mathbf{R}^2$ the qualitative theory of dynamical systems, as outlined in the previous chapter, appears to be particularly useful.

1.3. Conical streamlines, conical stagnation points

We will discuss some examples of conical flows and consider the corresponding topological structures in the cross flow plane Σ. Let the flow be described by a righthanded cartesian coordinate system x,y,z and let the velocity vector $\underline{V}$ be decomposed into its components u, v, w along the x, y and z axes, respectively.
The conical center will be taken at the origin (0,0,0) and the cross flow plane Σ is chosen normal to the x-axis. Since the flow is conical it is advantageous to use the conical coordinates $\eta = y/x$, $\zeta = z/x$ to describe the cross flow pattern in the plane Σ.
If the velocity vector $\underline{V} = u\,\underline{i} + v\,\underline{j} + w\,\underline{k}$ is decomposed nonorthogonally into a component $\underline{q}_r$ along the ray $\underline{r} = x\,\underline{i} + y\,\underline{j} + z\,\underline{k}$ and a component $\underline{q}_\Sigma$ in the plane Σ, then the latter defines the cross flow velocity vector.
The direction of $\underline{q}_\Sigma$ may be obtained as follows.

Because $\underline{q}_\Sigma$ lies in Σ it can be written as

$$\underline{q}_\Sigma = 0.\underline{i} + v_\Sigma \underline{j} + w_\Sigma \underline{k}$$

similarly the component along the ray is given by

$$\underline{q}_r = \frac{|\underline{q}_r|}{|\underline{r}|} \cdot (x\,\underline{i} + y\,\underline{j} + z\,\underline{k})$$

The cross flow components v_Σ and w_Σ follow from the requirement

$$\underline{V} = \underline{q}_\Sigma + \underline{q}_r$$

which yields

$$\frac{|\underline{q}_r|}{|\underline{r}|} = u/x$$

$$v_\Sigma = v - \frac{|\underline{q}_r|}{|\underline{r}|}\, y = v - u\eta$$

$$w_\Sigma = w - \frac{|\underline{q}_r|}{|\underline{r}|}\, z = w - u\zeta$$

This implies that the conical streamlines in Σ are solutions of the equation

$$\frac{d\eta}{d\zeta} = \frac{v - u\eta}{w - u\zeta}$$

or of the equivalent system

$$\dot{\eta} = v_\Sigma = v - u\eta\,, \qquad \dot{\zeta} = w_\Sigma = w - u\zeta \tag{2.1}$$

where the dot represents differentiation with respect to a parameter along the streamlines. Equations (2.1) will be considered as a dynamical system in $\mathbf{R}^2$ whose phase trajectories coincide with the conical streamlines in the cross flow plane. Conical stagnation points occur on those rays where the velocity vector is aligned with a local ray, yielding

$$\underline{q}_\Sigma = 0 \quad \text{or} \quad v - u\eta = 0; \qquad w - u\zeta = 0$$

Conical stagnation points may be identified with the singular points of system (2.1); they serve as the indispensable flow elements which may be used to compose more complicated conical flow patterns. In that way the knowledge about different types of conical stagnation points contribute to the understanding of the topology of conical flow patterns of increasing complexity.

To obtain a certain familiarity with the characteristic features of conical flow topology we proceed by discussing first the most simple example of a conical flow: the uniform parallel flow.

Consider a uniform parallel flow with a constant velocity u_∞ in the positive x-direction. This flow will be considered in the cross flow plane Σ. Since $u = u_\infty$, $v = w = 0$ the conical streamlines in Σ may be obtained from Eq. (2.1) as

$$\dot{\eta} = -u_\infty\eta \;; \qquad \dot{\zeta} = -u_\infty\zeta$$

This system has the general solution ζ = const. η. The conical streamlines are straight lines pointing to the origin where a conical stagnation point is formed, see Fig. 2.1.

For increasing time the conical streamlines are passed along in a direction indicated by the arrows if $u_\infty > 0$; for $u_\infty < 0$ this direction reverses.

The trace p and the Jacobian q are given by $p = -2u_\infty$ and $q = u_\infty^2$ indicating that the uniform flow parallel to the x-axis is represented as a starlike node in the cross flow plane Σ.

For $u_\infty > 0$ we refer to it as a conical sink, and for $u_\infty < 0$ as a conical source.

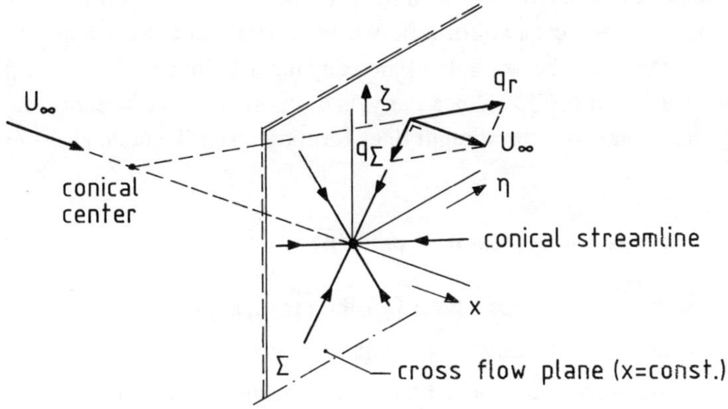

Fig. 2.1. Conical view of a supersonic flow $u = U_\infty$.

If the uniform flow is not aligned with the x-axis the starlike node may be found outside the origin. Since the starlike node itself remains unaltered it indicates that a uniform flow has a unique conical flow pattern without any regard to the specific orientation of the cross flow plane Σ. This conclusion may be illustrated as follows. Assume that a uniform flow, defined by $|\underline{v}|$ = constant, has velocity components u_∞, v_∞ and w_∞ in the x, y and z-direction, respectively. The conical streamlines satisfy the equation

$$\frac{d\eta}{d\zeta} = \frac{v_\infty - u_\infty \eta}{w_\infty - u_\infty \zeta}$$

having the solution $[\zeta - \frac{w_\infty}{u_\infty}] = \text{const.}\ [\eta - \frac{v_\infty}{u_\infty}]$.

The conical streamlines form a bundle of straight lines in Σ which point to a conical stagnation point at $\eta = \frac{v_\infty}{u_\infty}$, $\zeta = \frac{w_\infty}{u_\infty}$.

The conical stagnation point is a starlike node ($p = -2u_\infty$, $q = u_\infty^2$), its location $[\frac{v_\infty}{u_\infty}, \frac{w_\infty}{u_\infty}]$ coincides with the intersection of $\underline{V}$ with Σ

We have seen, irrespective of the orientation of the cross flow plane, that the uniform flow in the cross flow plane is uniquely represented by a starlike node. However the opposite that any starlike node in Σ reflects a uniform flow is in general not correct. This may be verified easily if one considers the example of an axisymmetric flow where every plane through the x-axis is a streamsurface. Since the velocity normal to the streamsurface is zero, the conical streamlines must be straight lines forming a starlike node at the origin (0,0).

1.4. Transition phenomena in conical flow patterns

Consider a circular cone with semi apex angle θ_c submerged, at incidence α, in a supersonic flow with free stream Mach number M_∞. The semi apex angle, incidence and free stream Mach number are such that a conical bow shock originates from the apex of the cone, see Fig. 2.2. The flow field will be described in a right handed cartesian coordinate system x,y,z such that the x-axis coincides with the cone axis and the flow field is symmetrical with respect to the x-z-plane. The cross flow plane Σ will be chosen normal to the x-axis at unit distance from the cone apex (conical center).

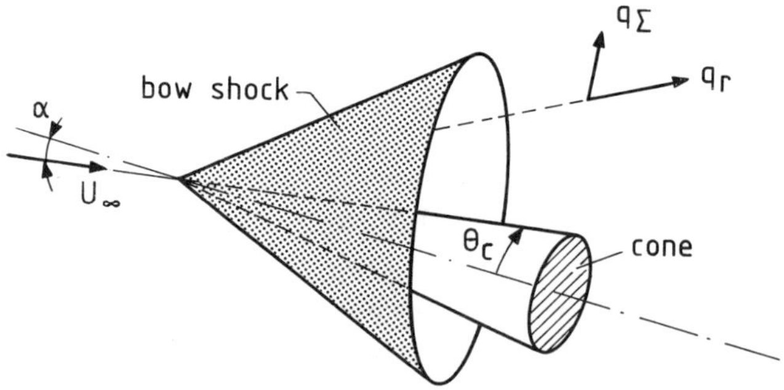

Fig. 2.2. Circular cone at incidence in supersonic flow with attached bow shock.

The boundary condition on the cone surface reads

$$(v - u\eta)\,\eta + (w - u\zeta)\,\zeta = 0$$

For $\alpha = 0$ the flow is axisymmetric having a bow wave with circular cross section in Σ. The conical streamlines outside and inside the bow wave are straight lines pointing to the origin (0,0).

Outside the bow wave the flow is uniform and inside of it we have a Taylor-Maccoll flow. Since the flow is axisymmetric with no velocity component normal to planes through the x-axis the cross flow components obey the relation $(v - u\eta)\,\zeta - (w - u\zeta)\,\eta = 0$. The boundary condition implies that on the cone surface both equations can be satisfied only if the cross flow vector $\underline{q}_{\Sigma} = (v - u\eta,\ w - u\zeta)$ vanishes identically. As a consequence the cone surface is the union of conical stagnation points which may be called a conical *stagnation line*.

All conical streamlines terminate perpendicularly at the conical stagnation line, see Fig. 2.3a.

If the incidence is slightly increased the topology of the flow pattern alters drastically; the conical streamlines encompass the cone surface and terminate at the leeside in a common point, see Fig. 2.3. At moderate incidence (say $\alpha/\theta_c \lessapprox 1$) and irrespective of the influence of viscosity, two conical stagnation points may be observed in the symmetry plane, one at the windward and another at the leeward generator of the cone. The conical stagnation point at the windward side is a saddle point, whereas at the leeward side a nodal point (vortical singularity) appears.

The evolution of the conical flow pattern due to a further increase of incidence depends on the influence of viscosity. For inviscid flow at high incidences the vortical singularity as already suggested by Ferri (1951) lifts off from the cone surface resulting in a saddle point at the leeward generator and a nodal point in the symmetry plane above the cone surface (Fig. 2.3d).

However, if viscosity is taken into account its influence tends to obscure the lift-off phenomenon, as is observed in the experiments made by Bannink & Nebbeling (1978). These experiments show that at high incidences flow separation leads to the generation of a vortex system on the leeward side of the cone (Fig. 2.3e). The separation point may be considered as the viscous analogue of a conical stagnation point having a saddle type structure with oblique separatrices. Similarly, the center of the vortex may be seen as a conical stagnation point with a spiral structure.

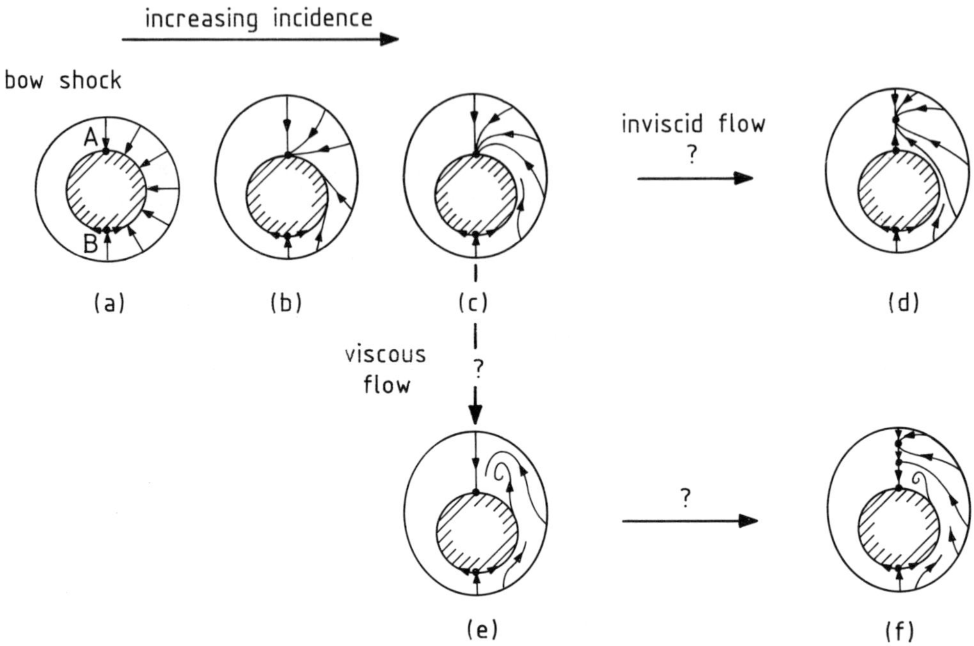

Fig. 2.3. Conical flow patterns for circular cone at incidence.

If the incidence is increased still further the spiral structure persists but the flow pattern above the vortex changes due to the appearance of a saddle and a node in the symmetry plane (Fig. 2.3f). If we compare the 'viscous' and the 'inviscid' route in the development of conical flow patterns we observe that viscosity can have a significant influence; it causes separation, vortex formation and a delay of the lift-off phenomenon.
Moreover, we observe in the 'viscous' as well as in the 'inviscid' scenario that the flow topology changes at certain combinations of the global parameters (incidence, Mach number, semi apex angle of the cone), suggesting that at the corresponding parameter values the conical flow pattern becomes structurally unstable and bifurcates.

Another example of changing flow topology may be found in the flow around an elliptical cone if the incidence is varied. A scenario of inviscid flow patterns for increasing incidence as obtained by numerical calculations (Grossman (1979)) is shown in Fig. 2.4. At zero incidence the flow pattern contains four conical stagnation points (two saddles and two nodes) on the body surface. The nodes N_1 and N_2 occur on the minor axis and the saddle points S_1 and S_2 are on the major axis of the ellips (Fig. 2.4a). Variation of incidence in the plane of the minor axis does not disturb the symmetry of the flow pattern with respect to this plane. For small incidences the saddle points S_1 and S_2 move towards the compression side of the cone whereas the nodal points N_1 and N_2 maintain their original positions on the minor axis (Fig. 2.4b). A further increase of incidence causes the coalescence of the saddle points S_1, S_2 and the node N_1. After coalescence a topologically different flow pattern results with two conical stagnation points at the cone surface, the original node N_2 at the leeward meridian and a saddle point S at the windward meridian (Fig. 2.4c). At the particular incidence where this transition occurs the conical flow pattern is structurally unstable and bifurcates.

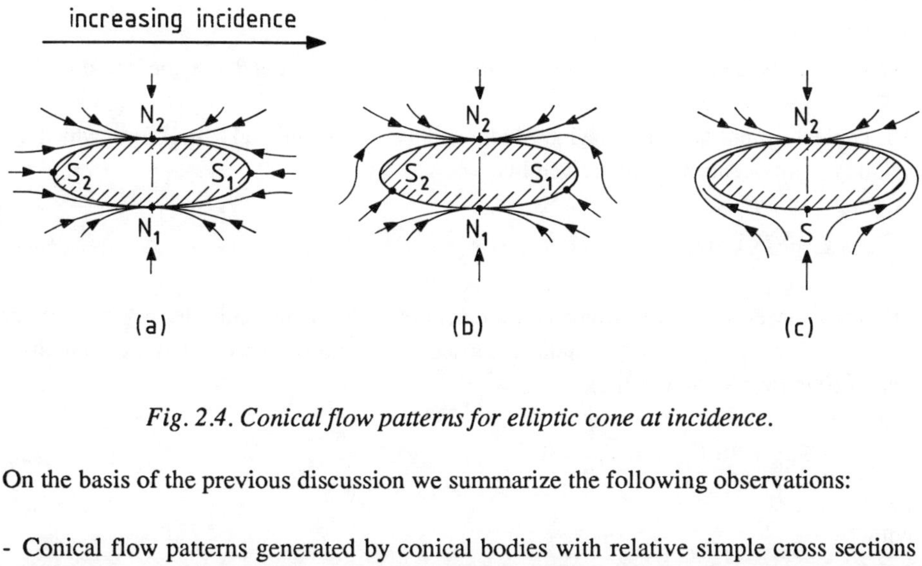

Fig. 2.4. Conical flow patterns for elliptic cone at incidence.

On the basis of the previous discussion we summarize the following observations:

- Conical flow patterns generated by conical bodies with relative simple cross sections may contain points where the cross flow velocity vanishes, the so called conical stagnation points. These points can occur on body contours as well as 'free' singularities in the flow field.

- Various types of flow patterns may be observed in the vicinity of conical stagnation points. If viscosity effects are important there appears, next to the more familiar orthogonal saddles and nodes, also oblique saddle points and vortices (foci?). Furthermore, a conical stagnation line or a starlike node can be encountered in special cases.

- Finally, there is a strong evidence that conical flow patterns may be structurally unstable leading to local bifurcation involving the generation or disappearance of conical stagnation points. Due to their sensitivity to perturbations the corresponding structural unstable flow patterns are hardly to realize in experiments or by numerical computations.

These observations motivate our systematic investigation of flow patterns near conical stagnation points as will be performed in the following paragraphs.

2. Local conical stagnation point solutions in irrotational flow

2.1. Conical potential equation

Consider the flow of an inviscid, non heat conducting, perfect gas with ratio of specific heats $\gamma = c_p/c_v$.

The flow is assumed to be irrotational so that a velocity potential Φ may be introduced by $\nabla \Phi = \underline{V}(u,v,w)$.

The conical similarity allows the introduction of a conical potential $F(\eta,\zeta)$ from which the velocity components may be derived by

$$\underline{V} = \nabla(x\, F(\eta,\zeta)) \tag{2.2}$$

From the conservation laws (mass and momentum) and the pressure-density relation for isentropic flow, it may be derived that F satisfies the nonlinear second order partial differential equation, Bulakh (1970)

$$A\, F_{\eta\eta} + 2B\, F_{\eta\zeta} + C\, F_{\zeta\zeta} = 0 \tag{2.3}$$

with

$$A = c^2(1+\eta^2) - (v-u\eta)^2$$

$$B = c^2\, \eta\zeta - (v-u\eta)(w-u\zeta)$$

$$C = c^2(1+\zeta^2) - (w-u\zeta)^2$$

where the speed of sound c is related to the velocity by

$$c^2 = \frac{\gamma-1}{2} (q_{max}^2 - u^2 - v^2 - w^2) \tag{2.4}$$

and q_{max} is the maximum speed, which we assume to be constant throughout the flow field. Equations (2.3) and (2.4) allow velocities to be nondimensionalized by q_{max}; as a result we put $q_{max} = 1$ so that (2.4) becomes

$$c^2 = \frac{\gamma-1}{2} (1 - u^2 - v^2 - w^2)$$

The velocity components u,v,w follow from (2.2) as

$$u = F - \eta F_\eta - \zeta F_\zeta$$

$$v = F_\eta$$

$$w = F_\zeta$$

Then the conical streamlines obey the equation

$$\dot{\eta} = v - u\eta = -\eta F + (1+\eta^2) F_\eta + \eta\zeta F_\zeta$$

$$\dot{\zeta} = w - u\zeta = -\zeta F + \eta\zeta F_\eta + (1+\zeta^2) F_\zeta$$

In the course of this study it appears also convenient to use polar coordinates $\eta = \rho \cos \varphi$, $\zeta = \rho \sin \varphi$ ($0 \leq \varphi < 2\pi$, $\rho > 0$) in the cross flow plane Σ; then the velocity components become

$$u = F - \rho F_\rho ,$$

$$v = F_\rho \cos \varphi - \frac{1}{\rho} F_\varphi \sin \varphi,$$

$$w = F_\rho \sin \varphi + \frac{1}{\rho} F_\varphi \cos \varphi$$

and (2.3) may be written as

$$\{c^2(1+\rho^2) - [\rho F-(1+\rho^2)F_\rho]^2\} F_{\rho\rho} + 2\{F-(\rho + \frac{1}{\rho}) F_\rho\} \{\frac{1}{\rho} F_{\rho\varphi} - \frac{1}{\rho^2} F_\varphi\} F_\varphi +$$

$$+ \{c^2 - \frac{F_\varphi^2}{\rho^2}\} \{\frac{F_{\varphi\varphi}}{\rho^2} + \frac{1}{\rho} F_\rho\} = 0 \qquad (2.5)$$

The conical streamlines are the integral curves of the system

$$\dot{\rho} = (1+\rho^2) F_\rho - \rho F$$

$$\dot{\varphi} = \frac{1}{\rho^2} F_\varphi$$

2.2. Conical stagnation point solutions

Approximate solutions of the conical potential equation (Eqs. 2.3, 2.5) near conical stagnation points will be found in the form of a series expansion in terms of the distance to the conical stagnation point. These solutions are locally valid and are first studied by Bakker, Bannink & Reyn (1981). The coordinate system is chosen with its origin in a conical stagnation point. The following series expansion of the conical potential is used

$$F = F_0\{1 + \rho^n F_n(\varphi) + \rho^m F_m(\varphi) + o(\rho^m)\}, \qquad 1 < n < m$$

where F_0 is a constant which equals the non-dimensionalized radial velocity component in the conical stagnation point.

The magnitude of F_0 has a significant influence on the conical streamline pattern. For supersonic conical stagnation points we have $c_0 < |F_0| < 1$, where c_0 is the speed of sound at the conical stagnation point. The solution $F = F_0$ yields a uniform parallel flow with a conical stagnation point at $\rho = 0$.

The condition $n < m$ provides $\rho^n F_n(\varphi)$ as the leading term in the expansion and the condition $1 < n$ assures that $\rho = 0$ is a conical stagnation point.

If Eq. (2.3) is evaluated in $\rho = 0$ (or $\eta = \zeta = 0$) the relation

$$F_{\eta\eta}(0,0) + F_{\zeta\zeta}(0,0) = 0$$

results which shows that near the conical stagnation point Eq. (2.3) is nearly the Laplace equation, which is also satisfied by the velocity potential Φ in incompressible plane flow. Therefore, the proposed expansion for F is suggested by *incompressible* plane stagnation point solutions which are given by

$$\Phi = a_n \rho^n \cos(n\varphi + \psi_n),\ n > 1$$

where a_n, ψ_n are constants. These solutions represent the well known flows in corners with opening angles $\omega = \pi/n$.

If the expansion for F is substituted into the conical potential equation, Eq. (2.5) and the results are ordered with respect to powers in ρ, the coefficient of the lowest-order term equated to zero yields

$$F_n'' + n^2 F_n = 0$$

with the solutions

$$F_n(\varphi) = \varepsilon_n \cos(n\varphi + \psi_n),\ n > 1 \tag{2.6}$$

where ε_n and ψ_n are arbitrary constants.
Thus, this term exactly equals the stagnation point solution for plane flow.
Since a freedom is left in the choice of the coordinate system, we may still rotate it around the x - axis such that $\psi_n = 0$.

When the next higher order terms are written out several cases for n and m have to be distinguished. The domain in the n,m-plane for which it is possible to determine the functions $F_n(\varphi)$ and $F_m(\varphi)$ is shown in Fig. 2.5.

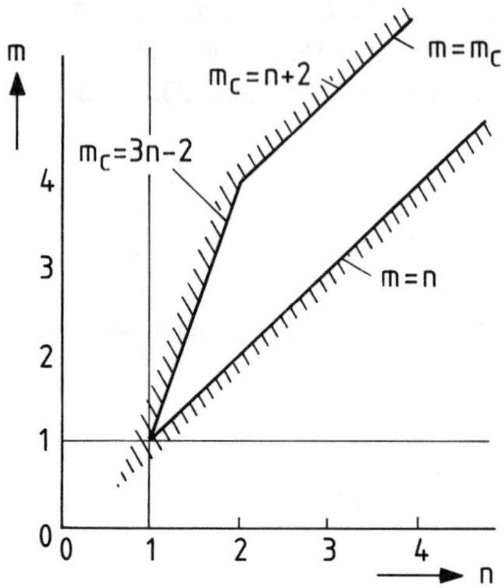

Fig. 2.5. Possible values of m and n for which $F_n(\varphi)$ and $F_m(\varphi)$ exist.

After equating the coefficient of the next highest order term to zero we obtain the following result

$$F''_m + m^2 F_m = 0 \ , \qquad n < m < m_c$$

$$F''_m + m^2 F_m = R_n(F_n, F'_n) \ , \qquad m = m_c$$

where R_n and m_c are specified, for different n, as follows:

$1 < n < 2$:	$R_n = n(n-1)\, M_o^2\, F_n\, (n^2 F_n{}^2 + F'_n{}^2)$	, $m_c = 3n - 2$
$n = 2$:	$R_n = -2F_2[1 - M_o^2\,(1-2F_2)^2] - 2M_o^2\, F_2'{}^2(1-F_2)$	, $m_c = 4$
$n > 2$:	$R_n = n\,(n-1)\, M_o^2\, F_n$	, $m_c = n + 2$

The quantity M_o is the Mach number in the conical stagnation point.

If $m > m_c$ we obtain $R_n = 0$ which yields, using (2.6), $F_n(\varphi) = 0$ implying that $m = m_c$ is an upper bound for m beyond which no solutions for $F_m(\varphi)$ exist.

If $F_m(\varphi)$ exists we find the following solutions

$$F_m(\varphi) = \beta_m \cos(m\varphi + \chi_m) , \qquad n < m < m_c$$

$$F_m(\varphi) = \beta_m \cos(m\varphi + \chi_m) + \gamma_m \cos(n\varphi) + \delta_m , \qquad m = m_c \tag{2.7}$$

The constants β_m and χ_m may be chosen arbitrarily, whereas the constants γ_m and δ_m are related to n, ε_n and M_o^2 as specified in the next table.

range of n	m_c	γ_m	δ_m
$1 < n < 2$	$3n - 2$	$\dfrac{n^3 M_o^2 \varepsilon_n^3}{4(2n-1)}$	0
$n = 2$	4	$\dfrac{\varepsilon_n}{6} (M_o^2 - 1 + 4\varepsilon_n^2 M_o^2)$	$\dfrac{-\varepsilon_n^2 M_o}{2}$
$n > 2$	$n + 2$	$\dfrac{n(n-1)\,(M_o^2-1)\,\varepsilon_n}{2(n+2)}$	0

Table: m_c, γ_m, δ_m

The solutions, discussed so far, are obtained for integer as well as for non integer values for n and/or m. For non integer values of one of these exponents, the solutions are not 2π-periodic; implying that it is impossible to fill out a full neighbourhood of the stagnation point with solutions such that the velocity is continuous. This implies that solutions which are not 2π-periodic have only physical relevance for conical stagnation points which are located on the contour of a conical body. Domains of multiplicity of the solutions are then masked by the interior of the body so that a single-valued flow field outside the body remains. In general this results in flow fields over partially smooth bodies with discontinuities in the slope where conical stagnation points occur.

With the aid of the listed solutions for the conical velocity potential the conical streamline pattern near conical stagnation points may now be determined.
In addition, the pressure distribution will be determined and compared with the streamline patterns.
In isentropic flows the pressure p is solely a function of the speed $|\underline{V}|$ and follows directly from Bernoulli's equation

$$\frac{\gamma}{\gamma-1}\,\frac{p}{\rho}+\frac{|\underline{V}|^2}{2}=\text{const.}$$

and Poisson's relation

$$p=\text{const.}\ \rho^{\gamma}$$

If the constants are evaluated in the conical stagnation point these equations may be written as

$$\left[\frac{p}{p_o}\right]^{\frac{\gamma-1}{\gamma}}=\frac{1-(u^2+v^2+w^2)}{1-u_o^2} \tag{2.8}$$

where zero subscripts refer to conditions in the conical stagnation point. Equation (2.8) shows that the isobars in the cross flow plane Σ coincide with constant velocity curves so that they can be determined from

$$\left(\frac{d\eta}{d\zeta}\right)_{p=\text{const.}}=\frac{\frac{\partial}{\partial\zeta}|\underline{V}|^2}{\frac{\partial}{\partial\eta}|\underline{V}|^2}\qquad\frac{-uu_\zeta-vv_\zeta-ww_\zeta}{uu_\eta+vv_\eta+ww_\eta}$$

Using $u=F-\eta v-\zeta w$ to calculate u_η and u_ζ, the isobar pattern is obtained from

$$\left(\frac{d\eta}{d\zeta}\right)_{p=\text{const.}}=\frac{-(v-u\eta)\,v_\zeta-(w-u\zeta)\,w_\zeta}{(v-u\eta)\,v_\eta+(w-u\zeta)\,w_\eta} \tag{2.9}$$

From this equation it may be observed that the isobar pattern in the cross flow plane Σ has a singularity in a point (η,ζ) if the following conditions are satisfied

$$(v-u\eta)\, v_\zeta + (w-u\zeta)\, w_\zeta = 0$$

$$(v-u\eta)\, v_\eta + (w-u\zeta)\, w_\eta = 0$$

thus either if

$$\underline{q}_\Sigma = 0 \quad \text{or} \quad \begin{vmatrix} v_\eta & v_\zeta \\ w_\eta & w_\zeta \end{vmatrix} = 0$$

This implies that a conical stagnation point always corresponds to a singularity in the isobar pattern. The reverse statement is not true, since if the second condition is satisfied (2.9) has a critical point but $\underline{q}_\Sigma$ is not equal to zero.

3. Classification of conical stagnation points in conical flow

3.1. First-order conical stagnation points

With the solutions, given in paragraph 2, we will investigate the three cases $1 < n < 2$, $n = 2$ and $n > 2$ separately. The discussion will be concentrated on the description of the resulting conical streamline patterns and related pressure distributions; details about the analysis may be found in the original publication of Bakker, Bannink & Reyn (1981).

<u>*Case $1 < n < 2$: oblique saddle points*</u>

Substitution of the solutions for $F_n(\varphi)$ and $F_m(\varphi)$ into the equation for the conical streamlines yields

$$\frac{d\rho}{d\varphi} = \frac{n\,\varepsilon_n\,\rho^{n+1}\cos n\varphi - \rho^3 + O(\rho^{m+1})}{-\,n\,\varepsilon_n\,\rho^n \sin n\varphi - \rho^3 + O(\rho^m)} \tag{2.10}$$

Introducing a parameter τ along the streamlines, we investigate this equation as a system in the (ρ,φ) plane. Then

$$\frac{d\rho}{d\tau} = n\,\varepsilon_n\,\rho\cos n\varphi - \rho^{3-n} + O(\rho^{m-n+1})$$

$$\frac{d\varphi}{d\tau} = -n\,\varepsilon_n \sin n\varphi + O(\rho^{m-n})$$

The singular points of this system on $\rho = 0$ are in $\varphi = \frac{k\pi}{n}$ (k = 0, ±1, ±2, ...) and the coefficient matrix of the locally linearized system has a trace $p = \pm n\ \varepsilon_n(1-n)$, and a Jacobian $q = -n^3 \varepsilon_n^2$. Since $q < 0$ the singular points are hyperbolic saddle points and the higher order terms do not change the saddle point character of these singular points. Therefore the lowest order terms are sufficient to determine the approximate shape of the conical streamlines:

$$\rho^n \sin n\varphi = C, \qquad 1 < n < 2$$

where C is constant along a streamline.
This equation represents the streamline pattern for a flow in a corner with an including angle $\omega = \pi/n$; thus $\pi/2 < \omega < \pi$ for $1 < n < 2$. The flow in the corner has a saddle point behaviour and the point is called an oblique saddle point; its separatrices form the including angle $\omega = \pi/n$ at the conical stagnation point. Since 2π is not an even multiple of π/n an oblique saddle point can occur only on a body contour with one or more streamlines in the solution that coincide with the body.
The pressure distribution near the conical stagnation point may be obtained from (2.8) as

$$\left(\frac{p}{p_o}\right)^{\frac{\gamma-1}{\gamma}} = 1 - \frac{\gamma-1}{2}\ M_o^2\ \varepsilon_n^2\ n^2\ \rho^{2n-2} + O(\rho^{n+m-2})$$

The pressure attains a maximum in the conical stagnation points and the isobars are to a first approximation concentric circles around the origin. In Fig. 2.6a the streamline pattern in a corner $0 \le \varphi \le \pi/n$ is shown. Fig. 2.6b gives the radial pressure distribution for various values of n, near an oblique saddle point. It appears that the pressure gradient at the conical stagnation point is infinite for corner angles $2\pi/3 < \omega < \pi$, corresponding to $1 < n < 3/2$, finite for $n = 3/2$ and zero for $3/2 < n < 2$.

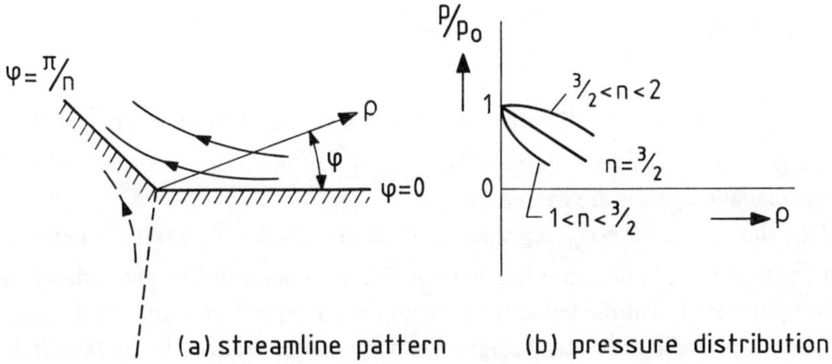

Fig. 2.6. Oblique saddle point (1 < n < 2).

<u>Case n = 2: saddle, nodes</u>

Substitution of the solutions $F_n(\varphi)$ and $F_m(\varphi)$ for n = 2 into the equation for the conical streamlines yields:

$$\frac{d\zeta}{d\eta} = \frac{-(2\varepsilon_2+1)\,\zeta + (mF_m \sin\varphi + F'_m \cos\varphi)\,\rho^{m-1} + \varepsilon_2\,\zeta(\eta^2-\zeta^2) + O(\rho^{m+1})}{(2\varepsilon_2-1)\,\eta\,(mF_m \cos\varphi - F'_m \sin\varphi)\,\rho^{m-1} + \varepsilon_2\,\eta(\eta^2-\zeta^2) + O(\rho^{m+1})} \tag{2.11}$$

with $2 < m \le 4$.

The eigenvalues of the coefficient matrix of the linear part: $\lambda_{1,2} = -1 \pm 2\varepsilon_2$ yield

$$p = \lambda_1 + \lambda_2 = -2, \qquad q = \lambda_1\lambda_2 = 1 - 4\varepsilon_2^2$$

If (2.11) is written in polar coordinates and considered in the (ρ, φ) plane, application of the theorem of Hartman-Grobmann (Chapter I) yields that the linearized system has the same streamline pattern as the nonlinear system if $|\varepsilon_2| \neq \frac{1}{2}$

In the η, ζ plane the linear system reads

$$\frac{d\eta}{d\tau} = (-1 + 2\varepsilon_2)\eta, \qquad \frac{d\zeta}{d\tau} = (-1 - 2\varepsilon_2)\zeta$$

which defines a continuous streamline pattern also in the η, ζ plane. For $|\varepsilon_2| = \frac{1}{2}$ a degenerate singular point occurs which will be investigated in section 3.3 where higher order conical stagnation points are treated.
Equation (2.7), shows that for non integer values of m the function $F_m(\varphi)$ is not periodic with period 2π; therefore, in that case it is not possible to cover a full neighbourhood of the stagnation point such that the velocity is continuous. If m = 3 or 4 and if all higher order terms are 2π-periodic the conical stagnation point can be realized in the flow field away from the body.

From the expressions: $p = -2$ and $q = 1 - 4\varepsilon_2^2$ we observe that conical stagnation points are represented in the p-q diagram, on that part of the line $p = -2$ where $p^2 \leq 4q$. It implies that in potential flow, conical stagnation points will be nodal points if $|\varepsilon_2| < \frac{1}{2}$ $(q > 0)$ and saddle points is if $|\varepsilon_2| > \frac{1}{2}$ $(q<0)$. In the special case $\varepsilon_2 = 0$ $(q = 1)$ the nodal point is a starlike node, with streamlines approaching the singularity in different directions. Sketches of these first order conical flow singularities are given in Fig. 2.7a; they may be observed in many practical flow situations such as the flow around cones, delta wings and axial corners.

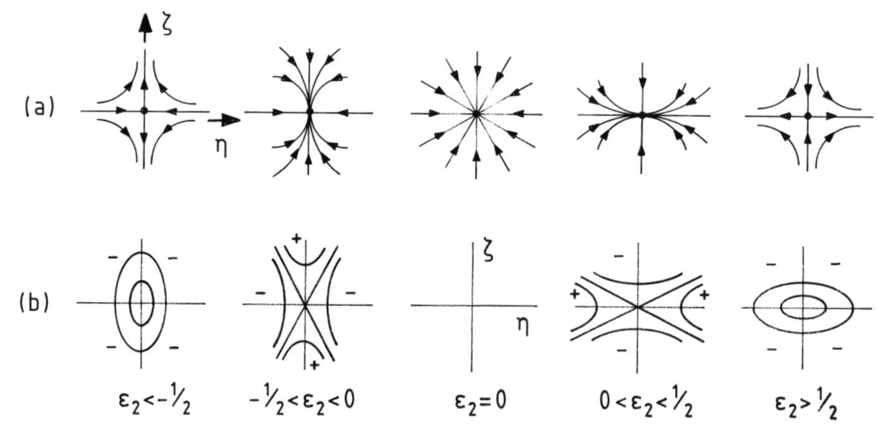

Fig. 2.7. Structurally stable conical flow patterns, n = 2, $|\varepsilon_2| = \frac{1}{2}$.
(a) conical streamlines, (b) isobars.

With Eq. (2.8) the pressure distribution becomes:

$$\left(\frac{p}{p_o}\right)^{\frac{\gamma-1}{\gamma}} = 1 - 2(\gamma\text{-}1)\, M_o^2\, \varepsilon_2 \left[(\varepsilon_2 - \tfrac{1}{2})\,\eta^2 + (\varepsilon_2 + \tfrac{1}{2})\,\zeta^2\right] + O(\rho^m) \tag{2.12}$$

If $|\varepsilon_2| > \frac{1}{2}$, and thus when a saddle point singularity of the streamlines occurs, the pressure attains a maximum in the conical stagnation point and the isobars are to a first approximation concentric ellipses around the origin.

If $0 < |\varepsilon_2| < \frac{1}{2}$ (nodal singularity) the isobars are to a first approximation nested hyperbolae with asymptotes given by

$$\zeta = \pm \left(\frac{\frac{1}{2} - \varepsilon_2}{\frac{1}{2} + \varepsilon_2} \right)^{\frac{1}{2}} \cdot \eta$$

In the region within the acute angle between the asymptotes the pressure is higher than in the conical stagnation point, whereas in the other regions the pressure is lower; see Fig. 2.7b. For $\varepsilon_2 = 0$ the pressure does not change to the order $m\gamma/(\gamma\text{-}1)$ of the distance from the conical stagnation points; this case actually corresponds to $n > 2$ which will be discussed hereafter.

Case n > 2: starlike nodes

Substitution of the solutions $F_n(\varphi)$ and $F_m(\varphi)$, as obtained for $n > 2$ into the equation for the conical streamlines yields

$$\frac{d\rho}{d\varphi} = \frac{-\rho^3 + n\,\varepsilon_n\,\rho^{n+1}\cos n\varphi + O(\rho^{m+1})}{-n\,\varepsilon_n\,\rho^n \sin n\varphi + O(\rho^{m+1})} \tag{2.13}$$

which, in fact, is equal to Eq. (2.10) for the case $1 < n < 2$.

Comparison of both equations however shows the decreasing influence of the ε_n term with respect to the radial velocity term $(-\rho^3)$ when $n > 2$.

Introducing the parameter τ along the streamlines we investigate (2.13) as a system in the (φ,σ) plane where $\sigma = \rho^{n-2}$ $(\sigma \geq 0)$.

Then

$$\dot{\varphi} = -n\,\varepsilon_n \sin n\,\varphi + O(\sigma^{\frac{m-n}{n-2}})$$

$$\dot{\sigma} = -(n-2) + n(n-2)\,\varepsilon_n\,\sigma \cos n\varphi + O(\sigma^{1+\frac{m-n}{n-2}})$$

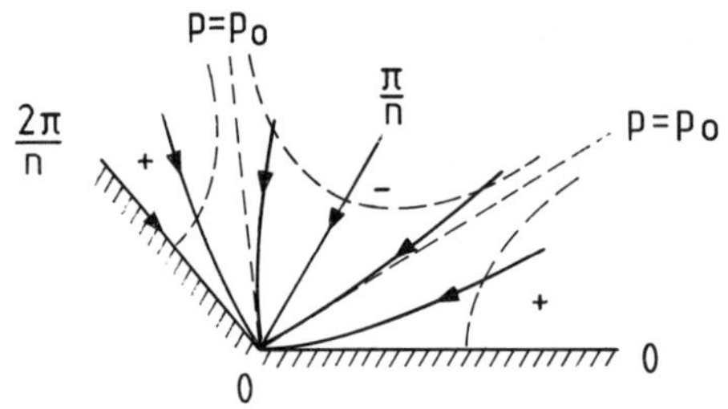

Fig. 2.8. Starlike node at n > 2.

The line $\sigma = 0$ contains only regular points, resulting in a single integral curve through any point of this line. Correspondingly,the streamlines approach the origin in the cross flow plane Σ from all directions. To the order considered, the lines $\varphi = k\,\pi/n$ ($k = 0, \pm1, \pm2, ..$) are conical streamlines dividing the flow field into sectors with an opening angle $\omega = \pi/n$ ($\omega < \pi/2$). The conical streamlines in each sector are curved at the singular point and the curvature is opposite in adjacent sectors. The flow in two adjacent sectors is shown in Fig. 2.8. Similar to the previous cases, this type of conical stagnation point in general cannot occur as a 'free' singularity in the flow field, certainly not if n is not an integer. The pressure distribution, which is given by the relation

$$\left(\frac{p}{p_0}\right)^{\frac{\gamma-1}{\gamma}} = 1 + (\gamma - 1)\,M_0^2\,(n-1)\,\varepsilon_n\,\rho^n \cos n\varphi + O(\rho^{2n-2})$$

shows a saddle behaviour near the conical stagnation point (see Fig. 2.8), with sectors correlated with the sectors in the streamline pattern.

A special example of this singularity is the conical stagnation point in a uniform flow. Then $\varepsilon_n = 0$ and the isobars shown in Fig. 2.8 do not apply since in this particular case the pressure is constant throughout the flow field.

The previous investigation of first-order conical stagnation point singularities shows us that in irrotational conical flow different types of singularities may be encountered such as oblique saddle points, saddles whose separatrices are perpendicular to each other, nodal points and starlike nodes in corners with an opening angle less than $\pi/2$ rad. However, other types that probably could be expected such as spiral points and oblique saddles in acute corners, as mentioned in paragraph 2.1, are not found in this classification. In this respect it must be emphasized that the results are obtained for irrotational flows; the extension of the classification for nonisentropic conical flows seems an interesting subject for further research.

3.2. Irrotational attachments and separations

It is of interest to discuss attachment and separation phenomenae in inviscid flows along a smooth conical body from a point of view of first order singularities. Such situations are likely to occur in many flow problems, especially in flows around bodies at high incidences.

Let us consider the details of the flow near a conical stagnation point which lies on the contour of a smooth conical body in such a flow. The smoothness of the contour implies that only conical stagnation points can occur which belong to the class n = 2.

Let the body contour in the cross flow plane Σ be defined by $\zeta = \zeta_c(\eta)$ and assume that the conical stagnation point is located on ζ_c at $\eta = \zeta = 0$. The body contour near the origin may be approximated by $\zeta_c = O(\eta^2)$.

Referring to the first-order singularities, obtained for n = 2, the following distinct streamline patterns may be considered.

$\varepsilon_2 > \frac{1}{2}$	a saddle point of attachment.
$0 < \varepsilon_2 < \frac{1}{2}$	a nodal point with an infinite number of streamlines tangent to the body contour (tangent node).
$\varepsilon_2 = 0$	a starlike node.
$-\frac{1}{2} < \varepsilon_2 < 0$	a nodal point having an infinite number of streamlines perpendicular to the body contour (orthogonal node).
$\varepsilon_2 < -\frac{1}{2}$	saddle point of separation.

Let us first give attention to the pressure distribution over the body surface in relation with the cross flow streamline pattern.

From Eq. (2.12), we obtain for the pressure over the surface:

$$\left(\frac{p}{p_o}\right)^{\frac{\gamma-1}{\gamma}} = 1 - 2(\gamma-1) M_o^2\, \varepsilon_2 (\varepsilon_2 - \tfrac{1}{2})\, \eta^2 + O(\eta^3)$$

This equation shows that the pressure has a *local minimum* if a tangent node occurs, whereas for an orthogonal node and a saddle point (attachment or separation) the pressure attains a *local maximum*. In the special case of a starlike node the pressure remains constant to second order.

Obviously only tangent nodes can be determined uniquely from a surface pressure distribution. To describe all possible streamline patterns in a unique sense the pressure behaviour perpendicular to the surface has to be known as well. A concise result may be obtained if the second order pressure derivatives are evaluated in the conical stagnation point. From Eq. (2.12) we obtain:

$$\hat{p}_{\eta\eta} = \frac{\partial^2 \hat{p}}{\partial \eta^2}(0,0) = 4\varepsilon_2(1 - 2\varepsilon_2)$$

$$\hat{p}_{\zeta\zeta} = \frac{\partial^2 \hat{p}}{\partial \zeta^2}(0,0) = -4\varepsilon_2(1 + 2\varepsilon_2)$$

where $\hat{p}$ is the pressure nondimensionalized by the dynamic pressure $\frac{1}{2}\gamma\, p_o\, M_o^2$ in the conical stagnation point, thus $\hat{p} = \dfrac{2p}{\gamma p_o M_o^2}$.

If ε_2 is eliminated the following relation between $\hat{p}_{\zeta\zeta}$ and $\hat{p}_{\eta\eta}$ results

$$(\hat{p}_{\zeta\zeta} - \hat{p}_{\eta\eta})^2 + 4(\hat{p}_{\zeta\zeta} + \hat{p}_{\eta\eta}) = 0$$

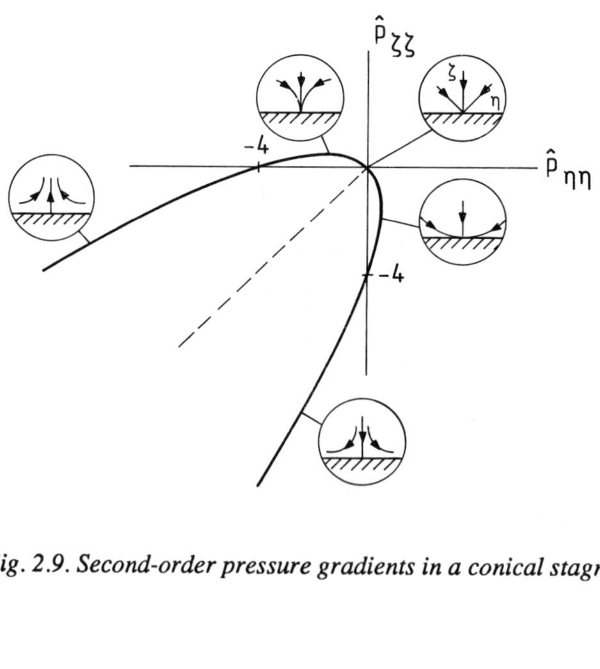

Fig. 2.9. Second-order pressure gradients in a conical stagnation point (n = 2).

From Fig. 2.9, which shows this relationship, it is easy to conclude that a *local maximum* of the surface pressure ($\hat{p}_{\eta\eta} < 0$) corresponds to:

- an orthogonal node if $\hat{p}_{\zeta\zeta} > 0$

- a saddle point of attachment if $\dfrac{\hat{p}_{\zeta\zeta}}{\hat{p}_{\eta\eta}} > 1$

- a saddle point of separation if $\dfrac{\hat{p}_{\zeta\zeta}}{\hat{p}_{\eta\eta}} < 1$

Furthermore it may be noticed that $\hat{p}_{\eta\eta}$ and $\hat{p}_{\zeta\zeta}$ never exceed the value $+\frac{1}{2}$ in isentropic conical flows.

The properties of conical stagnation points on a smooth body may also be studied in relation to the spatial inviscid streamline pattern on the body surface near those points. From the solution given by Eq. (2.6) the velocity components may be derived as

$$u = u_o \{1 - \varepsilon_2(\eta^2 - \zeta^2) + O(\rho^3)\}$$

$$v = 2u_o\, \varepsilon_2\, \eta + O(\rho^2)$$

$$w = -2u_o\, \varepsilon_2\, \zeta + O(\rho^2)$$

Since the body surface is approximately coincident with the plane $\zeta = 0$, the spatial surface streamlines follow from

$$\left[\frac{dy}{dx}\right]_{\zeta=0} = \frac{v}{u} = \frac{2\varepsilon_2\, \eta + O(\eta^2)}{1 + O(\eta^2)}$$

which may be integrated for small η to obtain the spatial streamline pattern close to $\eta = 0$ where zero cross flow appears. The correspondence of surface streamlines with cross flow streamline patterns is shown in Fig. 2.10.

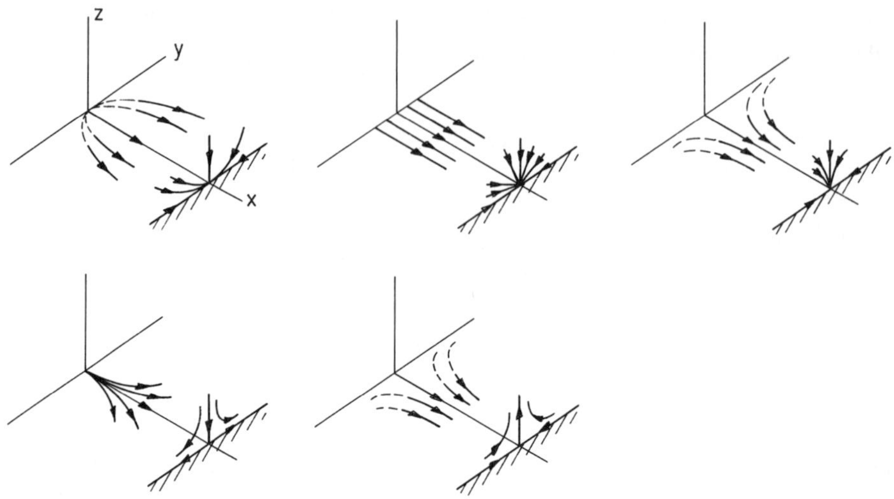

Fig. 2.10. Surface streamlines and corresponding cross flow pattern.

It appears for a *tangent node* that the surface streamlines, diverge from the attachment line (x-axis) forming a pattern with a so called 'gothic' behaviour. In the case of a *starlike node* the surface streamlines are to a first approximation parallel to each other.

The *orthogonal node* and the *saddle point of separation* generate spatial surface streamlines which converge to the ray with zero cross flow (x-axis). In the nodal case the rate of convergence is less than in the saddle case.
A *saddle point of attachment* is associated by a surface flow pattern diverging from the attachment line.

Finally it is of practical interest to obtain the correspondence of the cross flow streamline pattern with the spatial surface flow direction near the conical stagnation point.
The surface flow direction, denoted by ψ_s, is defined as the angle of the velocity vector with the local ray emanating from the conical center, see Fig. 2.11, then there is obtained

$$\tan \psi_s = \frac{v - u\eta}{u + v\eta}$$

The angle ψ_s passes through zero in a conical stagnation point. The rate at which ψ_s changes along the surface is determined by the particular cross flow pattern. Since

$$\frac{d\psi_s}{d\eta} = 2\,\varepsilon_2 - 1\,,$$

the cross flow pattern at a conical stagnation point is uniquely characterized by the directional gradient of surface streamlines (see Fig. 2.11).

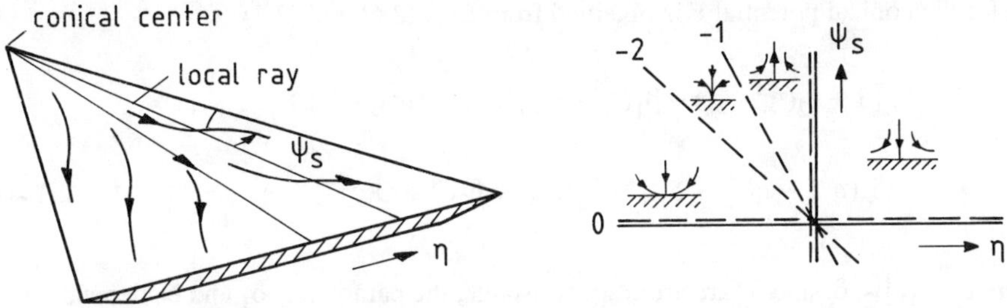

Fig. 2.11. Directional gradient of surface streamlines.

$$\frac{d\psi_s}{d\eta} > 0 \quad \text{: saddle point of attachment}$$

$$-1 < \frac{d\psi_s}{d\eta} < 0 \quad \text{: tangent node}$$

$$\frac{d\psi_s}{d\eta} = -1 \quad \text{: starlike node}$$

$$-2 < \frac{d\psi_s}{d\eta} < -1 \quad \text{: normal node}$$

$$\frac{d\psi_s}{d\eta} < -2 \quad \text{: saddle point of separation}$$

3.3. Higher-order conical stagnation points

Referring to section (3.1) of this paragraph it may be concluded that higher order conical stagnation points occur for n = 2 and $|\varepsilon_2| = \frac{1}{2}$. If $|\varepsilon_2| = \frac{1}{2}$ one of the eigenvalues of the locally linearized system is equal to zero and the nonlinear terms in Eq. (2.11) cannot be neglected when considering the streamline pattern.

The character of these higher order conical stagnation point singularities will now be investigated under the restriction that only 2π–periodic solutions of the conical potential are considered. Thus only those higher order singularities are discussed that can occur on smooth body contours or as 'free' singularities in the flow field. Then, the following solution for the conical potential F is obtained from Eqs. (2.6) and (2.7)

$$F = F_o\{1 + \varepsilon_2(\eta^2 - \zeta^2) + \beta_1(\eta^3 - 3\eta\zeta^2) - \beta_2(3\eta^2\zeta - \zeta^3) +$$

$$-2\delta_o(\eta^3\zeta - \eta\zeta^3) + \delta_1\eta^4 - \delta_2\eta^2\zeta^2 + \delta_3\zeta^4 + O(\rho^5)\} \qquad (2.14a)$$

where ε_2, β_1, β_2, δ_o and δ_2 are arbitrary constants; the parameters δ_1 and δ_3 depend on δ_2 by

$$\delta_1 = \frac{1}{6}\,\delta_2 + \frac{\varepsilon_2}{6}\,\{M_o^2\,(1 - 2\varepsilon_2)^2 - 1\}$$

$$\delta_3 = \frac{1}{6}\,\delta_2 - \frac{\varepsilon_2}{6}\,\{M_o^2\,(1 + 2\varepsilon_2)^2 - 1\} \qquad (2.14b)$$

and M_o is the local Mach number at the conical stagnation point, it depends on F_o as

$$M_o^2 = \frac{F_o^2}{c_o^2} = \frac{2}{\gamma-1}\,\frac{F_o^2}{1 - F_o^2}$$

The conical streamline pattern near conical stagnation points consists of the phase trajectories of the system that follows if Eq. (2.14) is substituted into the conical streamline equation $\dot{\eta} = v - u\eta$, $\dot{\zeta} = w - u\zeta$:

$$\dot{\eta} = v_\Sigma = (2\varepsilon_2 - 1)\eta + \frac{\mu_1}{2}(\eta^2 - \zeta^2) + \mu_2\eta\,\zeta + \frac{1}{6}\sum_0^3 \binom{3}{i} \nu_{i+1}\,\eta^{3-i}\zeta^i + O(\rho^4)$$

$$\dot{\zeta} = w_\Sigma = -(2\varepsilon_2+1)\zeta - \mu_1\,\eta\zeta + \frac{\mu_2}{2}(\eta^2 - \zeta^2) + \frac{1}{6}\sum_0^3 \binom{3}{i} \nu^*_{i+1}\,\zeta^{3-i}\eta^i + O(\rho^4) \tag{2.15}$$

The coefficients μ_i, ν_i and ν_i^* are new unknowns instead of β_i and δ_i. Their dependence on β_i and δ_i is irrelevant for the analysis. Of more importance are the mutual relations between ν_i and ν_i^* which follow from Eq. (2.14).

$$\begin{aligned} \nu_4^* &= -\nu_2^* = -\nu_4 = \nu_2 \\ \nu_1 + \nu_3 &= 4\varepsilon_2 M_o^2 (1 - 2\varepsilon_2)^2 \\ \nu_1^* + \nu_3^* &= -4\varepsilon_2 M_o^2 (1 + 2\varepsilon_2)^2 \\ \nu_1^* - \nu_1 &= -8\varepsilon_2 M_o^2 (1 + 4\varepsilon_2)^2 - 4\varepsilon_2 \end{aligned} \tag{2.16}$$

The flow patterns at $\varepsilon_2 = +\frac{1}{2}$ may be derived from those at $\varepsilon_2 = -\frac{1}{2}$ by performing the transformation

$$\begin{array}{ll} \eta \to -\zeta & \nu_1 \to \nu_1^* \\ \zeta \to \eta & \nu_1^* \to \nu_1 \\ \mu_1 \to \mu_2 & \nu_3 \to \nu_3^* \\ \mu_2 \to -\mu_1 & \nu_3^* \to \nu_3 \end{array}$$

which shows that the nature of the degenerate singularities at $\varepsilon_2 = +\frac{1}{2}$ and at $\varepsilon_2 = -\frac{1}{2}$ is similar and that a rotation of $\pi/2$ rad transforms them into each other. Therefore, we restrict the discussion on higher order singularities to one value of ε_2 only, say $\varepsilon_2 = -\frac{1}{2}$.

For $\varepsilon_2 = -\frac{1}{2}$, system (2.15) has a degenerate singularity at (0,0) with one zero eigenvalue. Then Andronov's Theorem 65 about this type of multiple singular points may be applied (c.f. Chapter I) from which follows that $\dot{\eta} = 0$ on a curve which is approximated by

$$\eta = f(\zeta) = -\frac{\mu_1}{4}\zeta^2 + O(\zeta^3)$$

and that on this curve

$$\dot{\zeta} = \Delta_m\,\zeta^m + O(\zeta^{m+1}) = -\frac{\mu_2}{2}\zeta^2 + [\frac{\nu_1^*}{6} + \frac{\mu_1^2}{4}]\,\zeta^3 + O(\zeta^4) \qquad (2.17)$$

According to Andronov different degenerate singularities can be distinguished now.

Case: $\mu_2 \neq 0$, saddle-node

For $\mu_2 \neq 0$ there follows $m = 2$ and $\Delta_m = -\mu_2/2$ so that the multiple singular point is a saddle-node, it has one parabolic sector (nodal type) and two hyperbolic sectors (saddle type), see Fig. 2.12a.

An unlimited number of streamlines, constituting the nodal part, approach the conical stagnation point along the direction $\eta = 0$ whereas two single streamlines approach the saddle-node along the direction $\zeta = 0$; they are convex towards the nodal part of the saddle-node. This implies that if a saddle-node is fitted on a body surface, the nodal part occurs for convex surfaces whereas for concave surfaces only the saddle part appears.

The particular streamline which separates the parabolic sector from the hyperbolic sectors has the approximate form

$$\zeta = -\frac{1}{8}\mu_2\eta^2 + O(\eta^3)$$

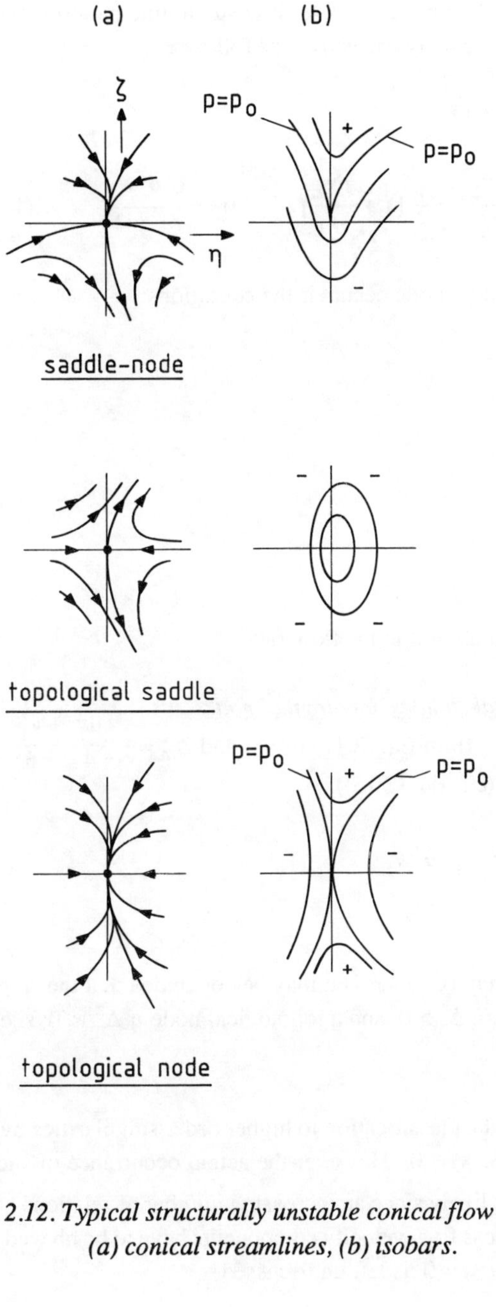

Fig. 2.12. Typical structurally unstable conical flow patterns. (a) conical streamlines, (b) isobars.

The occurrence of a saddle-node in cross flow streamline patterns may be expressed in terms of the cross flow velocity derivatives as follows.

From Eq. (2.15) we obtain

$$\varepsilon_2 = -\frac{1}{2}\left[1 + \frac{\partial w_\Sigma}{\partial \zeta}\right] = \frac{1}{2}\left[1 + \frac{\partial v_\Sigma}{\partial \eta}\right], \qquad \mu_2 = -\frac{\partial^2 w_\Sigma}{\partial \zeta^2}$$

which implies that a saddle node occurs if the conditions:

1. $v_\Sigma = w_\Sigma = 0,$

2. $\dfrac{\partial w_\Sigma}{\partial \zeta} = 0 \; (\varepsilon_2 = -\frac{1}{2})$

3. $\dfrac{\partial^2 w_\Sigma}{\partial \zeta^2} \neq 0$

are satisfied in a common point in the flow field.

Case $\mu_2 = 0$; topological saddles, topological nodes

For $\mu_2 = 0$ there follows from Eq. (2.17) m = 3 and $\Delta_m = [\frac{\mu_1^2}{4} + \frac{v_1^*}{6}]$ or, expressed in cross flow derivatives, (c.f. Eq. (2.15)).

$$\Delta_m = \frac{1}{4}\left(\frac{\partial^2 w_\Sigma}{\partial \eta\, \partial \zeta}\right)^2 + \frac{1}{6}\frac{\partial^3 w_\Sigma}{\partial \zeta^3}$$

From Andronov's theorem (Chapter I) it may be concluded that the singular point is a topological saddle point if $\Delta_m > 0$ and a topological node if $\Delta_m < 0$; sketches of them are shown in Fig. 2.12a.

It is possible to extend the classification to higher order singularities by investigating the particular case $\Delta_m(m = 3) = 0$. However the actual occurrence of such singularities in real flows becomes less likely since an increasing number of relations between higher order derivatives of the cross flow velocity components have to be obeyed by the flow. Therefore the case $\Delta_m(m = 3) = 0$ is left undiscussed.

The isobar pattern near the higher order singularities ($\varepsilon_2 = -\frac{1}{2}$), discussed so far, may be obtained by substituting Eq. (2.14) into Eq. (2.9). The isobars are then the integral curves of the following system

$$\frac{d\eta}{d\tau} = \frac{3}{4}\mu_2\eta^2 - \frac{1}{2}\mu_1\eta\zeta + \frac{\mu^2}{4}\zeta^2 - \sum_0^3 A_i\eta^{3-i}\zeta^i$$

$$\frac{d\zeta}{d\tau} = \eta - \frac{5}{4}\mu_1\eta^2 - \frac{5}{4}\mu_2\eta\zeta + \frac{1}{4}\mu_1\zeta^2 + \sum_0^3 B_i\eta^{3-i}\zeta^i$$

where τ is a parameter along an isobar, A_i and B_i are coefficients which depend on μ_i, ν_i and ν_i^*. They will be specified in the course of the analysis if their respective values are needed. Both eigenvalues of the linearized system near (0,0) are equal to zero so that Andronov's theorem about this type of multiple singular points may be applied (c.f. Chapter I). The curve on which $\frac{d\zeta}{d\tau} = 0$ is

$$\eta = -\frac{\mu_1}{4}\zeta^2 + O(\zeta^3)$$

and on this curve

$$\frac{d\eta}{d\tau} = \Delta_k\zeta^k + O(\zeta^{k+1}) = \frac{\mu_2^2}{4}\zeta^2 + \left(\frac{\mu_1^2}{8} - A_3\right)\zeta^3 + O(\zeta^4)$$

where

$$A_3 = \frac{\nu_1^*}{12} + \frac{\mu_1^2 + \mu_2^2}{4}$$

Hence it follows that for $\mu_2 \neq 0$, $k = 2$ which implies that the isobars form a cusp near the singularity. As we have shown before, the case $\mu_2 \neq 0$ corresponds to a saddle-node of conical streamlines: the isobar pattern is sketched in Fig. 2.12b.
For $\mu_2 = 0$, there is obtained

$$k = 3 \text{ and } \Delta_k = \frac{\mu_1^2}{8} - A_3 = -\frac{1}{2}\Delta_m$$

implying that the isobar pattern exhibits a center for $\Delta_k < 0$ and a topological saddle for $\Delta_k > 0$.

Since Δ_k and Δ_m have opposite sign we obtain as a final result that a topological node of conical streamlines $(\Delta_m < 0)$ corresponds to a topological saddle of isobars $(\Delta_k > 0)$ and a topological saddle point of conical streamlines $(\Delta_m > 0)$ gives a center point of isobars $(\Delta_k < 0)$; sketches of these pressure figures are shown in Fig. 2.12b. In these pictures, regions with a pressure exceeding that in the stagnation point are indicated by a plus sign, whereas a lower pressure is denoted by a minus sign.

4. Analytical unfoldings in conical flows

4.1. Bifurcation parameters

The second-order system Eq. (2.15), which describes the behaviour of conical streamlines near conical stagnation points, contains several unknown parameters $(\varepsilon_2, \mu_1, \nu_i, \nu_i^*)$ which define a multiple parameter family of functions in the right-hand side of the system. These functions are represented by Taylor series wherein these parameters may be identified as spatial derivatives of the cross flow velocity components in the conical stagnation point.

The topological structure near a conical stagnation point will depend on the actual parameter values of $\varepsilon_2, \mu_i, \nu_i, \nu_i^*$ which can also act as bifurcation parameters.

Since higher-order stagnation points occur at $|\varepsilon_2| = \frac{1}{2}$ (which are saddle-nodes if $\mu_2 \neq 0$ and topological nodes or topological saddles if $\mu_2 = 0$), the bifurcation parameters are obviously ε_2 and μ_2 with bifurcation values $|\varepsilon_2| = \frac{1}{2}$ and $\mu_2 = 0$, respectively. The right hand side of Eq. (2.15) is represented by a Taylor series, therefore we consider only bifurcations caused by analytic perturbations.

The bifurcation of a saddle node is described by at least one parameter (ε_2) whereas the bifurcation of a topological node and a topological saddle point is governed by at least two parameters (ε_2 and μ_2). So far the parameters in Eq. (2.15) are recognized which may cause bifurcation. However, the bifurcation processes that can be treated now are restric-

ted to flow patterns which contain after bifurcation at least one conical stagnation point. This property of the bifurcation process is due by the fact that Taylor expansions are performed with respect to a conical stagnation point so that the system has always a singular point at the origin (0,0). This restriction is rather unfortunate, because it prevents the description of spontaneous generation of conical stagnation points; furthermore it prevents also the observation of flow phenomenae where two or more singularities merge together and disappear after coalescence.
This difficulty may be circumvented if the bifurcation of degenerate singularities allows for nonzero cross flow at the original location of the higher order singularity. This means that the magnitude of the cross flow velocity $|\underline{q}_\Sigma|$ at $\eta = \zeta = 0$ has to be taken as an additional independent bifurcation parameter. Together with ε_2 and μ_2 we expect a *codimension-two* bifurcation for the saddle-node and a *codimension-three* bifurcation for the topological node and the topological saddle.

For a proper description of these bifurcations the system of Eq. (2.15) is not appropriate since the third bifurcation parameter $|\underline{q}_\Sigma|$ does not appear in the right hand side of (2.15). A modified formulation follows if in Eq. (2.1) conical flow solutions are used which are obtained by expanding $F(\eta, \zeta)$ near a regular point in the flow field.
In general this expansion may be performed near the origin $\eta = \zeta = 0$ so that $F(\eta, \zeta)$ may be approximated by:

$$F = F_o \left[1 + \sum_{n=1}^{N} \rho^n F_n(\varphi) + O(\rho^{N+1})\right] \tag{2.18}$$

Among the unknown functions $F_n(\varphi)$, (n = 1, .. N) (which will be solved up till N = 4 in the next paragraph), only $F_1(\varphi)$ contributes to the cross flow velocity $\underline{q}_\Sigma$ at the origin, since:

$$|\underline{q}_\Sigma|^2 (0,0) = F_o^2 \left\{F_1^2 + \left[\frac{dF_1}{d\varphi}\right]^2\right\} \tag{2.19}$$

4.2. Approximate solution near regular points

To construct approximate solutions of the conical potential equation, Eq. (2.5) near points with nonvanishing cross flow and to be in accordance with stagnation point solution developed in section 2.2, we will expand the conical potential in terms of the distance to the origin $\eta = \zeta = 0$ by the power series:

$$F = F_o[1 + \sum_{n=1}^{N} \rho^n F_n(\varphi) + O(\rho^{N+1})] \tag{2.18}$$

If this equation is substituted into Eq. (2.5) and the result is ordered in powers of ρ, ordinary differential equations for the functions $F_n(\varphi)$ are obtained.

The equation for $F_1(\varphi)$ gives

$$\frac{\gamma-1}{2}\{1 - F_o^2 - F_o^2[F_1^2 + \frac{\gamma+1}{\gamma-1}F_1'^2]\}\{F_1'' + F_1\} = 0$$

which will be satisfied by two different types of solutions

$$F_1^{(1)}(\varphi) = a_1 \cos(\varphi - \psi_1) \tag{2.20}$$

and

$$F_1^{(2)}(\varphi) = \frac{\sqrt{1-F_o^2}}{F_o} \sin[\sqrt{\frac{\gamma-1}{\gamma+1}}\varphi - \varphi_o] \tag{2.21}$$

The second solution $F_1^{(2)}(\varphi)$ is not 2π-periodic and should therefore be omitted here. The solution describes the conical analogue of a Prandtl-Meyer fan and represents a possible flow near a discontinuity in the slope of a conical body.

The first solution $F_1^{(1)}(\varphi)$ contains two arbitrary constants a_1 and ψ_1, which may identified with the magnitude and direction of the cross flow velocity at the origin, respectively. The relation of $|q_\Sigma|$ with a_1 is given by:

$$|q_\Sigma|(0,0) = F_o\sqrt{F_1^2 + F_1'^2} = F_o \cdot a_1 \tag{2.22}$$

The quantity a_1, which is the bifurcation parameter we are looking for, will also appear in the solutions for $F_2(\varphi)$, $F_3(\varphi)$ and $F_4(\varphi)$, cf. Bakker (1984).

Since a_1 has the bifurcation value $a_1 = 0$, only solutions for small a_1 appear to be significant.

The laborious calculations in obtaining these solutions, have been made by Bakker (1984), here we suffice to list the results

$$F_2(\varphi) = \varepsilon_2\{\kappa \cos \psi_2 + \cos(2\varphi - 2\psi_1 + \psi_2)\}$$

where κ is a function of the cross flow Mach number at the origin $M_c = \dfrac{|q_\Sigma|}{c(0,0)}$

$$\kappa = \frac{a_1^2 \; F_o^2}{(\gamma\text{-}1)\,(1\text{-}F_o^2) - \gamma\, a_1^2 \; F_o^2} = \frac{M_c^2}{2 - M_c^2} \tag{2.23}$$

and ε_2 and ψ_2 are arbitrary constants.

We may use the freedom, still existing in the choice of the coordinate system to rotate it around the x-axis in such a way that $\psi_2 - 2\,\psi_1 = 0$ resulting into

$$F_2(\varphi) = \varepsilon_2\{\kappa \cos 2\psi_1 + \cos 2\varphi\} \tag{2.24}$$

Then the solution for $F_3(\varphi)$ and $F_4(\varphi)$ are found to be

$$F_3(\varphi) = c_{31} \sin 3\varphi + c_{32} \cos 3\varphi + \frac{a_1\, \varepsilon_2\, F_o^2}{(\gamma\text{-}1)(1\text{-}F_o^2)} \{(2\varepsilon_2\text{-}1) \cos \varphi \cos \psi_1 +$$

$$+ (2\varepsilon_2\text{+}1) \sin \varphi \sin \psi_1\} + O(a_1^2) \tag{2.25}$$

and

$$F_4(\varphi) = c_{41} \sin 4\varphi + c_{42} \cos 4\varphi - M_c^2 \,\frac{1}{2}\, \varepsilon_2^2 \; +$$

$$+ \frac{1}{6}\,\varepsilon_2\{(1 + 4\varepsilon_2^2)\, M_c^2 \; - 1\} \cos 2\varphi + O(a_1) \tag{2.26}$$

where c_{31}, c_{32}, c_{41} and c_{42} are arbitrary constants.

Note that the functions $F_3(\varphi)$ and $F_4(\varphi)$ are specified to the order $O(a_1^2)$ and $O(a_1)$ respectively; the specification of higher order terms has been omitted here because their influence on the bifurcation process appears to be immaterial.
The system defining the conical streamline pattern may now be written as

$$\begin{aligned}\dot{\eta} = P(\eta,\zeta) = F_o\{a_1 \cos\psi_1 + (2\bar{\varepsilon}_2 - 1)\eta + \bar{\mu}_1 \frac{\eta^2}{2} + \bar{\mu}_2 \eta\zeta + \bar{\mu}_3 \frac{\zeta^2}{2} + \\ + \frac{1}{6}\sum_0^3 [{}^3_i]\,\bar{\nu}_{i+1}\,\eta^{3-i}\zeta^i + O(\rho^4)\\[1em] \dot{\zeta} = Q(\eta,\zeta) = F_o\{a_1 \sin\psi_1 - (2\bar{\bar{\varepsilon}}_2 + 1)\zeta + \bar{\mu}_2 \frac{\eta^2}{2} + \bar{\mu}_3 \eta\zeta + \bar{\mu}_4 \frac{\zeta^2}{2} + \\ + \frac{1}{6}\sum_0^3 [{}^3_i]\,\bar{\nu}^*_{i+1}\,\zeta^{3-i}\eta^i + O(\rho^4)\end{aligned} \tag{2.27}$$

where

$$\bar{\varepsilon}_2 = \varepsilon_2(1 + \tfrac{1}{2}\, a_1^2\, M_o^2 \cos 2\psi_1) + O(a_1^3)$$

$$\bar{\bar{\varepsilon}}_2 = \varepsilon_2(1 - \tfrac{1}{2} a_1^2\, M_o^2 \cos 2\psi_1) + O(a_1^3)$$

$$\bar{\mu}_1 = \mu_1 + 3a_1\,\varepsilon_2(2\varepsilon_2 - 1)\, M_o^2 \cos\psi_1 + O(a_1^2)$$

$$\bar{\mu}_2 = \mu_2 + a_1\,\varepsilon_2(2\varepsilon_2 + 1)\, M_o^2 \sin\psi_1 + O(a_1^2)$$

$$\bar{\mu}_3 = -\mu_1 + a_1\,\varepsilon_2(2\varepsilon_2 - 1)\, M_o^2 \cos\psi_1 + O(a_1^2)$$

$$\bar{\mu}_4 = -\mu_2 + 3a_1\,\varepsilon_2(2\varepsilon_2 + 1)\, M_o^2 \sin\psi_1 + O(a_1^2)$$

The coefficients $\bar{\nu}_i$ and $\bar{\nu}^*_i$ are related to ν_i and ν^*_i as $\bar{\nu}_i = \nu_i + O(a_1)$ and $\bar{\nu}^*_i = \nu^*_i + O(a_1)$ respectively.
Note that Eq. (2.27) reduces to Eq. (2.15) when taking $a_1 = 0$; representing the case that the origin becomes a conical stagnation point.

4.3. Saddle-node bifurcation

A saddle-node occurs in a conical flow pattern if the cross flow velocity components v_Σ and w_Σ satisfy the following conditions:

1. $v_\Sigma = w_\Sigma = 0 \qquad (a_1 = 0)$

2. $\dfrac{\partial w_\Sigma}{\partial \zeta} = 0 \qquad (\varepsilon_2 = -\tfrac{1}{2})$

3. $\dfrac{\partial^2 w_\Sigma}{\partial \zeta^2} \neq 0 \qquad (\mu_2 \neq 0)$

The conical streamline pattern near this degenerate singularity may be obtained from (2.27) yielding

$$\begin{aligned} \dot{\eta} &= -2\eta + \mu_1 \frac{\eta^2}{2} + \mu_2 \eta\zeta - \mu_1 \frac{\zeta^2}{2} + O(\rho^3) \\ \dot{\zeta} &= +\mu_2 \frac{\eta^2}{2} - \mu_1 \eta\zeta - \mu_2 \frac{\zeta^2}{2} + O(\rho^3) \end{aligned} \tag{2.28}$$

where the immaterial (positive) constant F_o has been used to scale the parameter along the conical streamline.
Bifurcation of the saddle-node occurs if the parameters ε_2 and a_1 are perturbed with respect to their respective bifurcation values $\varepsilon_2 = -\frac{1}{2}$ and $a_1 = 0$. Assume that these perturbations are given by ε and α which are small compared to unity, $\varepsilon_2 = -\frac{1}{2} + \varepsilon$, $a_1 = 0 + \alpha$.
The unfolding of Eq. (2.28) which satisfies the conical flow equations may be derived from Eq. (2.27) by substituting $\varepsilon 2 = -\frac{1}{2} + \varepsilon$ and $a_1 = \alpha$.
The qualitative behaviour of the flow pattern occuring after bifurcation is mainly determined by the neighbouring conical stagnation points $P_i(\eta_i, \zeta_i)$ that might appear after bifurcation. To obtain the locations of these points for small α and ε we use center manifold theory (Chapter I).

For that, system (2.27) is extended with the equation $\dot{\alpha} = 0$ in order to obtain a system (in $\mathbf{R}^3$) that contains ε as the only parameter; α is now considered as an additional independent variable.

For $\varepsilon = 0$ this extended system becomes:

$$\dot{\zeta} = \alpha \sin \psi_1 \qquad + \mu_2 \frac{\eta^2}{2} - \mu_1 \eta\zeta - \mu_2 \frac{\zeta^2}{2} + O(3)$$

$$\dot{\alpha} = 0$$

$$\dot{\eta} = \alpha \cos \psi_1 - 2\eta \qquad + \mu_1 \frac{\eta^2}{2} + \mu_2 \eta\zeta - \mu_1 \frac{\zeta^2}{2} + O(3)$$

which has a degenerate singularity at (0,0,0); O(3) denotes third and higher order terms composed of ζ, α and η.

The influence of ε is analyzed in the center manifold that is tangent to the center eigenspace of:

$$\begin{pmatrix} 0 & \sin \psi_1 & 0 \\ 0 & 0 & 0 \\ 0 & \cos \psi_1 & -2 \end{pmatrix} \qquad \text{with eigenvalues } \lambda_1 = \lambda_2 = 0,\ \lambda_3 = -2$$

The corresponding eigenspaces are

$$E^c = (0,2,\cos\psi_1)\,x(1,0,0), \qquad E^s = (0,0,1), \qquad E^u = \phi$$

The center manifold W^c : $\eta = h(\zeta,\alpha)$, tangent to E^c satisfies the equation

$$\begin{pmatrix} \frac{\partial h}{\partial \zeta} & \frac{\partial h}{\partial \alpha} \\ 0 & 0 \end{pmatrix} \begin{pmatrix} \dot{\zeta} \\ \dot{\alpha} \end{pmatrix} - \begin{pmatrix} \dot{\eta} \\ 0 \end{pmatrix} = 0$$

with the boundary conditions: $h(0,0) = h_\zeta(0,0) = 0$, $h_\alpha(0,0) = \frac{1}{2} \cos \psi_1$.

The approximation of the center manifold near (0,0,0) becomes

$$W^c : h(\zeta,\alpha) = h_1\alpha + h_2\alpha^2 + h_3\alpha\zeta + h_4\zeta^2 + O(3)$$

where O(3) denotes terms of $O(\zeta^3, \zeta^2\alpha, \zeta\alpha^2, \alpha^3)$.

The coefficients h_1, h_2, h_3 and h_4 are determined as

$$h_1 = \frac{1}{2}\cos\psi_1$$

$$h_2 = \frac{\mu_1}{16}\cos^2\psi_1 - \frac{\mu_2}{8}\sin\psi_1\cos\psi_1 - \frac{\mu_1}{8}\sin^2\psi_1$$

$$h_3 = (\mu_1\sin\psi_1 + \mu_2\cos\psi_1)/4$$

$$h_4 = -\frac{\mu_1}{4}$$

The essential features of the unfolding follow if (2.27) is projected on W^c:

$$\dot{\zeta} = \alpha\sin\psi_1 + \frac{\mu_2}{8}\alpha^2\cos^2(\psi_1) - [2\varepsilon - \frac{\mu_1\alpha}{2}\cos\psi_1]\,\zeta - \mu_2\frac{\zeta^2}{2} + O(3) \qquad (2.29)$$

which shows that for $\alpha = 0$ and $\varepsilon = 0$ the ζ^2-term is the lowest-order nonvanishing term implying a codimension-one bifurcation so that only one perturbation parameter should be necessary to decribe the bifurcation in a persistent way. However, Eq. (2.29) also shows that the physical interpretation of this bifurcation needs the use of two parameters (α and ε). The locations of the neighbouring conical stagnation points will depend on these parameters and may be obtained from the expression for W^c and the projected system on W^c.

For small values of α and ε the approximate location is given by the leading terms:

$$\zeta_i = \frac{-2\varepsilon \pm 2\sqrt{\varepsilon^2 + \frac{1}{2}\mu_2\alpha\sin\psi_1}}{\mu_2}$$

$$\eta_i = \frac{1}{2}\alpha\cos\psi_1 - \frac{1}{4}\mu_1\zeta_i^2 \qquad (2.30a)$$

if $\sin\psi_1 \neq 0$, $i = 1,2$, and

$$\zeta_i = \frac{-(4\varepsilon + \mu_1\alpha) \pm \sqrt{(4\varepsilon + \mu_1\alpha)^2 + (\mu_2\alpha)^2}}{2\mu_2}$$

$$\eta_i = \frac{1}{2}\alpha \qquad (2.30b)$$

if $\sin\psi_1 = 0$ ($\cos\psi_1 = +1$), i = 1,2.

If $\sin\psi_1 \neq 0$, the saddle-node falls apart into two conical stagnation points if the quantity $\mu_2\,\alpha\sin\psi_1$, which we will call $\bar{\alpha}$, exceeds the value $-2\ \varepsilon^2$. The type of these points follows by evaluating the trace p and the Jacobian q at these points; it appears that the plus sign in Eq. (2.30a) corresponds to a stable node and the minus sign to a saddle point. For $\bar{\alpha} = -2\ \varepsilon^2$ a nongeneric bifurcation occurs; the saddle-node remains present and is shifted with respect to its original position. If $\bar{\alpha} < -2\ \varepsilon^2$ the saddle-node disappears without generating any neighbouring singularity.
A qualitative impression of saddle-node bifurcation in conical flow is given in Fig. 2.13 which shows the bifurcation set together with conical flow patterns.

If $\sin\psi_1 = 0$ saddle-node bifurcation will occur under the forced condition that the cross flow velocity component w_Σ remains zero. As a consequence, always two conical stagnation points will appear after bifurcation: a saddle-point and a stable node.

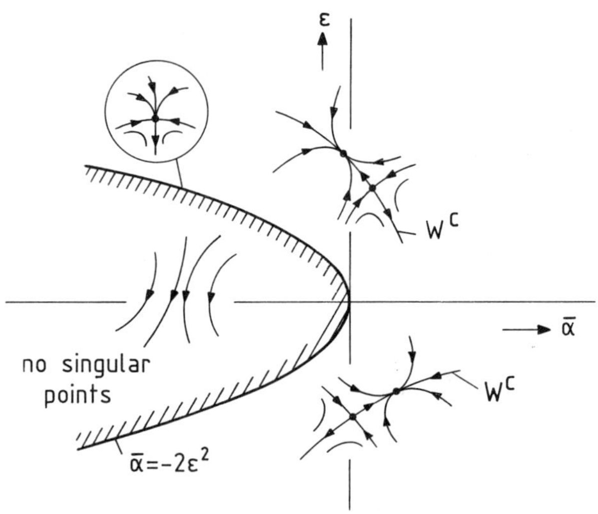

Fig. 2.13. Bifurcation of the saddle-node.

Let us now discuss further the case sin $\psi_1 \neq 0$ and see how saddle-node bifurcation may be used to explain some transition phenomena in conical flows. The bifurcation set, Fig. 2.13, shows that a saddle-node may bifurcate into two different ways. The first possibility is its falling apart into two structurally stable singularities and the other is its disappearance leaving no singularities at all. The existence of these two bifurcation modes is the reason for two different transition processes in conical flow patterns. Each of them may be identified with a traverse along a particular curve in the $\bar{\alpha}$, ε-diagram. The first transition phenomenon occurs if one moves for example along $\bar{\alpha}=0$ from $\varepsilon<0$ to $\varepsilon>0$. The transition is marked by the fact that conical stagnation points remain present during the transition. If ε increases from $\varepsilon < 0$ to $\varepsilon>0$, a saddle and a node move together and coalesce into the saddle-node ($\varepsilon = 0$); after coalescence the saddle and the node appear again and move away from each other. This type of bifurcation, which will be called: 'saddle-node bifurcation of the first type', explains the 'lift-off' phenomenon which was observed at the leeward side of a cone at sufficiently high incidence (this chapter, paragraph 1.4). At lift off, the conical stagnation point at the body surface in the leeward singularity plane would be a saddle-node; the nodal part is formed by the streamlines outside the body surface and the saddle part is found if the flow is extended inside the body (Bakker & Bannink 1974). A decrease of incidence leaves a nodal point on the cone surface and creates a saddle point in the solution inside the cone. If the incidence is increased beyond the value where lift-off occurs the saddle-node falls apart into a saddle point attached to the cone surface and a node moving away from the body.

A different type of transition exists if we move in the bifurcation set (Fig. 2.13) from $\bar{\alpha}<0$ to $\bar{\alpha}>0$ along $\varepsilon=0$. We observe that a regular flow field exists as long as $\bar{\alpha}<0$ and that only at $\bar{\alpha}=0$ a saddle-node appears. A further increase to $\bar{\alpha}>0$ resuls into the formation of a saddle point and a stable node.
In conclusion, this transition proces is characterized by the simultaneous appearance of two conical stagnation points the 'birth' of which is caused by the sudden occurrence of a degenerate saddle-node. This example of saddle-node bifurcation will be called: 'saddle-node bifurcation of the second type'. It explains the observations of Bannink & Nebbeling (1978) that a saddle and a node suddely appear in the flow around a circular cone beyond a particular high incidence.

4.4. Bifurcation of topological saddle point

A topological saddle point occurs in a conical flow pattern if the cross flow components v_Σ and w_Σ obey the following conditions

1. $v_\Sigma = w_\Sigma = 0 \; (a_1 = 0)$

2. $$\frac{\partial w_\Sigma}{\partial \zeta} = 0 \; (\varepsilon_2 = -\tfrac{1}{2})$$

3. $$\frac{\partial^2 w_\Sigma}{\partial \zeta^2} = 0 \; (\mu_2 = 0)$$

4. $$\Delta_m = \frac{1}{4}\left(\frac{\partial^2 w_\Sigma}{\partial \eta \partial \zeta}\right)^2 + \frac{1}{6}\frac{\partial^3 w_\Sigma}{\partial \zeta^3} > 0$$

The conical streamline pattern near this degenerate singularity may be obtained from Eq. (2.27) as:

$$\begin{aligned} \dot{\eta} &= v_\Sigma = -2\eta + \frac{\mu_1}{2}(\eta^2 - \zeta^2) + O(\rho^3) \\ \dot{\zeta} &= w_\Sigma = -\mu_1 \eta\zeta + v_1^* \frac{\zeta^3}{6} + O(\rho^3) \end{aligned} \tag{2.31}$$

where the immaterial (positive) constant F_o is included in the parametrization.

On the right-hand side, the term $v_1^* \frac{\zeta^3}{6}$ is written explicitly whereas the remaining third- and higher-order terms are denoted as $O(\rho^3)$ since their contribution to the qualitative flow picture appears to be a higher order effect.

The detailed structure of the topological saddle point depends on the actual values of μ_1 and v_1^*. The topological saddle point has four separatrices, two of them approach to the conical stagnation point along the η-axis and the others along the ζ-direction. The latter are curved and can be approximated by the parabola $\eta = -\frac{1}{4}\mu_1 \zeta^2$.

Bifurcation of the topological saddle point occurs if the bifurcation parameters ε_2, a_1 and μ_2 are perturbed with respect to their respective bifurcation values $\varepsilon_2 = -\frac{1}{2}$, $a_1 = 0$,

$\mu_2 = 0$. Analogous to the saddle-node case we assume these perturbations to be given by ε, α and μ which are small compared to unity; $\varepsilon_2 = -\frac{1}{2} + \varepsilon$, $a_1 = 0 + \alpha$, $\mu_2 = 0 + \mu$.
The unfolding of Eq. (2.31) satisfying the conical flow equations may be derived from Eq. (2.27) by substituting the perturbation parameters ε, α and μ.
To obtain the locations of neighbouring conical stagnation points $P_i(\eta_i, \zeta_i)$ as a function of the perturbation parameters: α, ε and μ we use center manifold theory. For that system (2.27) is extended with the equation $\dot{\alpha} = 0$ in order to obtain a system (in $\mathbf{R}^3$) that contains ε and μ as the parameters, α is now considered as an additional independent variable.
For $\varepsilon = \mu = 0$ the extended system reads:

$$\dot{\zeta} = \alpha \sin \psi_1 - \mu_1 \eta\zeta + O(3)$$

$$\dot{\alpha} = 0$$

$$\dot{\eta} = \alpha \cos \psi_1 - 2\eta + \mu_1 \frac{\eta^2}{2} + \mu_1 \frac{\zeta^2}{2} + O(3)$$

The influence of ε and μ will be analyzed in the center manifold that is tangent to the center eigenspace of

$$\begin{pmatrix} 0 & \sin \psi_1 & 0 \\ 0 & 0 & 0 \\ 0 & \cos \psi_1 & -2 \end{pmatrix}$$

The eigenspaces of the linear system are

$$E^c = (0,2, \cos \psi_1) \ x(1,0,0), \ \ E^s = (0,0,1), \ \ E^u = \phi$$

The center manifold $W^c : \eta = h(\zeta,\alpha)$ with boundary conditions $h(0,0) = h_\zeta(0,0) = 0$, $h_\alpha(0,0) = \frac{1}{2} \cos \psi_1$ may be approximated near (0,0,0) as

$$h(\zeta,\alpha) = \frac{1}{2} \cos \psi_1 \alpha + O(\alpha^2) + \frac{\mu_1}{4} \sin \psi_1 \alpha\zeta - \frac{\mu_1}{4} \zeta^2 + O(3)$$

The essential features of the unfolding of (2.31) follow by projecting (2.27) on the center manifold W^c : for $\sin\psi_1 \neq 0$ the unfolding on W^c becomes

$$\dot{\zeta} = \alpha \sin\psi_1 - 2\varepsilon\zeta - \frac{1}{2}\mu\zeta^2 + \Delta_m \zeta^3 \tag{2.32}$$

with $\Delta_m = \dfrac{\mu_1^2}{4} + \dfrac{\nu_1^*}{6}$ and where the redundant higher-order terms in α, ε and μ are dropped out. Equation (2.32) shows that the projected system undergoes codimension-two bifurcation; however, the physical interpretation of this bifurcation is described by the three parameters ε, α and μ. The neighbouring conical stagnation points $P_i(\eta_i, \zeta_i)$ which lie on W^c will depend on these parameters and may be obtained from

$$\eta_i = -\frac{\mu_1}{4}\zeta_i^2$$

$$\alpha \sin\psi_1 - 2\varepsilon\zeta_i - \frac{1}{2}\mu\zeta_i^2 + \Delta_m\zeta_i^3 = 0, \quad i = 1,2,3$$

Since the ζ-coordinate of P_i satisfies a cubic equation, there are one, two or three real solutions depending on the sign of D:

$$D = -(\hat{\varepsilon} + \hat{\mu}^2)^3 + (\hat{\mu}^3 + \frac{3}{2}\hat{\varepsilon}\hat{\mu} - \hat{\alpha})^2 \tag{2.33}$$

where we have introduced the shorthand notation

$$\hat{\alpha} = \frac{\alpha \sin\psi_1}{2\Delta_m}, \quad \hat{\varepsilon} = \frac{2}{3}\frac{\varepsilon}{\Delta_m}, \quad \text{and } \hat{\mu} = \frac{\mu}{6\Delta_m}$$

If $D < 0$, there are three real and different solutions; for $D = 0$ again three real solutions appear but at least two of them are equal; finally, if $D > 0$ one real solution appears and two are conjugate complex. The bifurcation set of the topological saddle point consists of the surface $D = 0$ bordering domains in $\mathbf{R}^3$ where the flow topology is different. A view of this surface is given in Fig. 2.14.

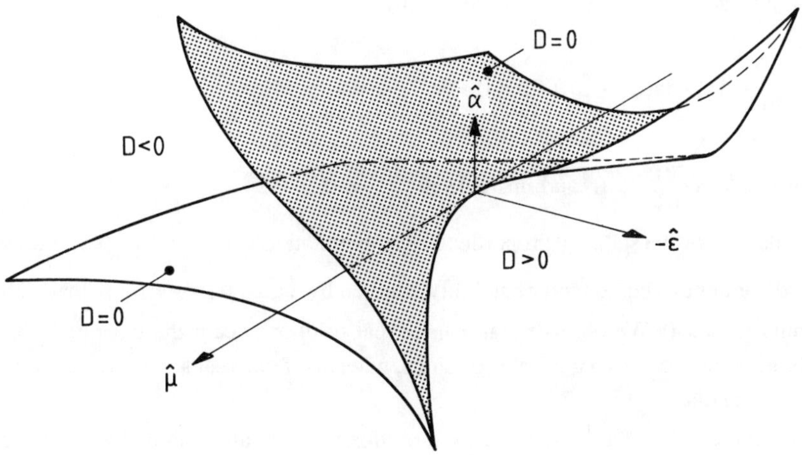

Fig. 2.14. Bifurcation set D = 0. $D = (\hat{\mu}^3 + \frac{3}{2}\hat{\varepsilon}\hat{\mu} - \hat{\alpha})^2 - (\hat{\varepsilon} + \hat{\mu}^2)^3$.

The surface $D = 0$ has two branches which terminate on the curve $\hat{\varepsilon} = -\hat{\mu}^2$, $\hat{\alpha} = -\frac{1}{2}\hat{\mu}^3$ where cross sections of $D = 0$ with $\hat{\mu}$ = constant form a cusp. If $\hat{\mu} = 0$ we observe that a bifurcation solution with three conical stagnation points occurs in a 'wedge' type region near the positive $\hat{\varepsilon}$-axis. For $\hat{\mu} \neq 0$ this region deformes more or less, so that also three conical stagnation points can occur for an appropriate negative value of $\hat{\varepsilon}$. At the cusp point the two solutions on $D = 0$ coalesce into the triple solution: $\zeta_i = \hat{\mu}$.

To obtain the resulting conical flow patterns it suffices to analyze the bifurcation on the center manifold W^c and to investigate the stability properties of the occurring singularities.
The flow on W^c which is given by Eq. (2.32) may be recast in the form

$$\dot{\zeta} = f(\zeta; \hat{\alpha}, \hat{\varepsilon}, \hat{\mu}) = \Delta_m\{(\zeta - \hat{\mu})^3 - 3a(\zeta - \hat{\mu}) + 2b\}$$

where $a = \hat{\mu}^2 + \hat{\varepsilon}$ and $b = \hat{\alpha} - \frac{3}{2}\hat{\varepsilon}\hat{\mu} - \hat{\mu}^3$.

Singular points $\zeta = \zeta_i$ on W^c satisfy the relation

$$(\zeta_i - \hat{\mu})^3 - 3a(\zeta_i - \hat{\mu}) + 2b = 0 \;;$$

they are stable if $\frac{df}{d\zeta} < 0$ and unstable if $\frac{df}{d\zeta} > 0$.

For the analysis of the stability properties it is convenient to consider $\zeta_i - \hat{\mu}$ as a function of b for different a. The domain of stability is given by $|\zeta_i - \hat{\mu}| < \sqrt{a}$ so that stability exists only for $a > 0$. We observe that if three real solutions occur there is one stable solution lying in between two unstable solutions, whereas if one real solution occurs it must be an unstable one.

In the special case $b = \pm a\sqrt{a}$ $(D = 0)$ two solutions appear, an unstable hyperbolic point and a nonhyperbolic singularity having a stable and an unstable neighbourhood. In the very special case $a = b = 0$ a nonhyperbolic unstable singularity occurs.

The flow near the center manifold in the cross flow plane may be constructed easily by remembering that the unstable eigenspace E^u is empty. It implies that stable singular points on W^c correspond to stable nodes whereas unstable singular points reveal saddle points in the cross flow plane.

The nonhyperbolic singularity on W^c having a stable and an unstable neighbourhood represents a saddle-node.

In Fig. 2.15 qualitative sketches are given of these cross flow streamline patterns near the center manifold W^c.

The main results about topological saddle point bifurcation may be summarized as follows:

- A topological saddle point bifurcates into a stable nodal point and two stable saddle points if $D < 0$.
- If $D > 0$ the topological saddle point is transformed into a single structurally stable saddle point.
- For $D = 0$ a nongeneric bifurcation occurs which generates a saddle point and a saddle-node. In the very special case $D = 0$ and $\hat{\varepsilon} = -\hat{\mu}^2$, $\hat{\alpha} = -\frac{1}{2}\hat{\mu}^3$ $(a = b = 0)$ the topological saddle point remains present and is shifted along W^c with respect to its original position.

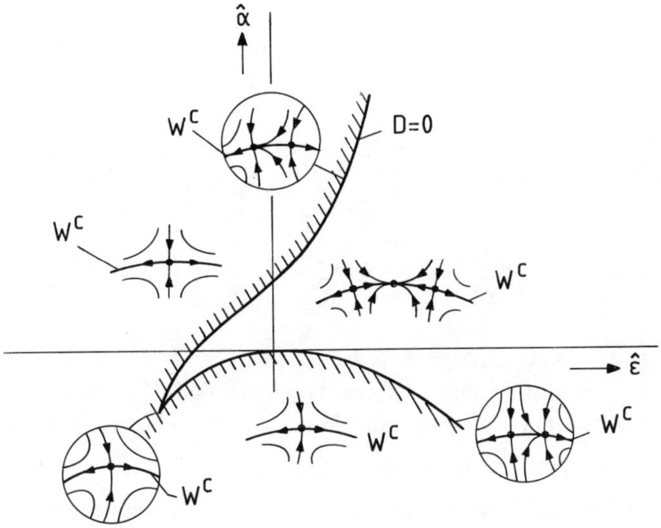

Fig. 2.15. Bifurcation of topological saddle ($\hat{\mu} > 0$) bifurcation set ($D = 0$) and flow patterns near center manifold W^c.

An example of saddle point bifurcation may be encountered in the flow past an elliptic cone at incidence. Cross flow streamline patterns at various incidences were already given in Fig. 2.4. Beyond a particular incidence (say $\alpha = \alpha_c$) the flow structure at the windward side changes qualitatively when the saddle points S_1, S_2 and the nodal point N_1 are replaced by a single saddle point S. The transition that occurs may be explained as saddle point bifurcation occurring at $\alpha = \alpha_c$. The topological saddle point appears at the particular incidence $\alpha = \alpha_c$. A small decrease of incidence causes bifurcation into a stable node N_1 flanked by two saddle points S_1 and S_2.

If the incidence is increased beyond α_c the topological saddle point changes into a saddle point S.
Our knowledge about saddle point bifurcation will be applied to show some details of this flow phenomenon. First we note that the flow is symmetric with respect to the minor axis and, moreover, there is always a conical stagnation point present where this axis intersects

the body contour.
This implies that the flow near the windward side may be described by the system (see Eq. (2.27))

$$\dot{\eta} = v_\Sigma = (-2 + 2\varepsilon)\,\eta + \frac{\mu_1}{2}\,(\eta^2 - \zeta^2) + \nu_1\,\frac{\eta^3}{6} + \frac{\nu_3}{2}\,\eta\zeta^2 + O(\rho^4)$$

$$\dot{\zeta} = w_\Sigma = -2\varepsilon\zeta - \mu_1\,\eta\zeta + \frac{\nu_3^*}{2}\,\eta^2\zeta + \nu_1^*\,\frac{\zeta^3}{6} + O(\rho^4) \tag{2.34}$$

where the bifurcation parameters α and μ are omitted.
The η-axis coincides with the minor axis of the elliptic body (positive inwards) and the conical stagnation point at the windward side is taken at the origin. Near the origin the body contour will be approximated by $\eta = k_1\,\zeta^2 + k_2\,\zeta^4 + ...$ where k_1 and k_2 are positive quantities which depend on the principal vertex angles of the elliptic cone.
The boundary condition that the body contour coincides with a conical streamline is satisfied if

$$\mu_1 = 4k_1\,(1 - 3\varepsilon) + O(\varepsilon^2)$$

Since ε is the only bifurcation parameter, saddle point bifurcation occurs at $\varepsilon = 0$.
Details of the bifurcation process may be observed clearly from the behaviour of the curves $v_\Sigma = 0$ and $w_\Sigma = 0$ if ε varies near zero.
Apart from the symmetry plane ($\zeta = 0$) where $w_\Sigma = 0$, these curves will be given by the approximations (see Eq. (2.34))

$$v_\Sigma = 0: \quad \eta = k_1(1 - 2\varepsilon)\,\zeta^2 + (k_2 + \Delta_m\,k_1)\,\zeta^4 + \;...$$

$$w_\Sigma = 0: \quad \eta = \frac{\varepsilon}{2k_1} - (\Delta_m - 4k_1^2\,)\,\frac{\zeta^2}{4k_1} + ...$$

The influence of ε on the curves $v_\Sigma = 0$ and $w_\Sigma = 0$ are sketched in Fig. 2.16 for $\Delta_m > 0$.
For $\varepsilon < 0$ these curves do not intersect at all, meaning that no conical stagnation points exist outside the symmetry plane. In the symmetry plane where $w_\Sigma = 0$ there is a saddle point at the body surface, corresponding to the flow situation beyond the critical angle $\alpha > \alpha_c$.

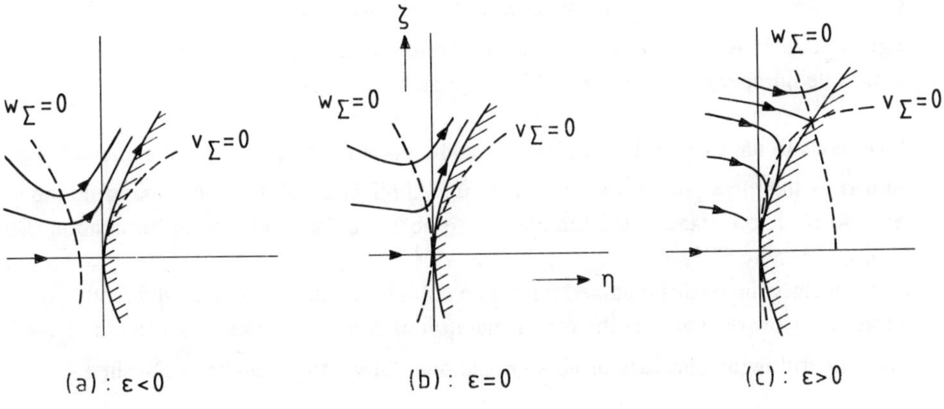

Fig. 2.16. Flow near windward side of elliptic cone.

If ε is increased the curve $w_\Sigma = 0$ shifts in positive η-direction and becomes tangent to the curve $v_\Sigma = 0$ at the origin if $\varepsilon = 0$; in this case we have to do with the structurally unstable flow at $\alpha = \alpha_c$.

If ε is increased beyond $\varepsilon = 0$, the $w_\Sigma = 0$ curve moves further and intersects the $v_\Sigma = 0$ curve on the body contour at $\zeta^2 = \frac{2\varepsilon}{\Delta_m}$. Thus for $\varepsilon > 0$ two conical stagnation points appear at the body surface outside the symmetry plane.

4.5. Bifurcation of topological node

In conical flow a topological node occurs if the cross flow components v_Σ and w_Σ obey conditions similar to those given for the topological saddle point (previous section) except that the sign of Δ_m has changed.

The conical streamline pattern near this degenerate singularity may be obtained from Eq. (2.27) as

$$\dot{\eta} = v_\Sigma = -2\eta + \frac{\mu_1}{2}(\eta^2 - \zeta^2) + O(\rho^3)$$

$$\dot{\zeta} = w_\Sigma = -\mu_1 \eta\zeta + v_1^* \frac{\zeta^3}{2} + O(\rho^3)$$

The detailed structure of the flow depends on the actual values of μ_1 and ν_1^*. The topological node contains two streamlines tending to the stagnation point in η-direction and an infinite number of streamlines which approach in the ζ-direction.

Analogous to the case of the topological saddle point the bifurcation phenomenon is governed by the three parameters $\hat{\alpha}$, $\hat{\varepsilon}$ and $\hat{\mu}$. Application of the center manifold theory leads to equations exactly the same as those obtained for saddle point bifurcation (Eq. (2.32)).
Consequently the results obtained for saddle point bifurcation can be used directly. To interpret the flow on and near the center manifold it suffices to take into account $\Delta_m < 0$. Then the following conclusions about nodal point bifurcation can be established.

Nodal point bifurcation in conical flow is governed by the local quantity D (Eq. (2.33)) as follows:

- For $D < 0$, a topological node bifurcates into a saddle-point flanked by two nodal points.
- For $D > 0$ the topological node transforms into a hyperbolic nodal point.
- For $D = 0$ a nongeneric bifurcation occurs generating a nodal point and a structurally unstable saddle-node. In the very special case $D = 0$ and $\hat{\varepsilon} = -\hat{\mu}^2$, $\hat{\alpha} = -\frac{1}{2}\hat{\mu}^3$ the topological node remains as such and is shifted along the path $\eta = -\frac{\mu_1}{4}\zeta^2$.

A qualitative impression of the flow at and near the center manifold may be obtained by consulting Fig. 2.17.

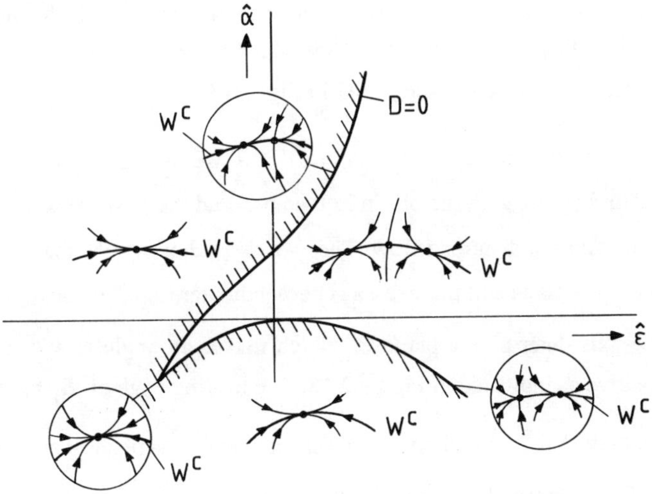

Fig. 2.17. Bifurcation of topological node ($\hat{\mu} > 0$) bifurcation set ($D = 0$) and flow patterns near center manifold W^c.

5. External corner flow; a nonanalytical unfolding of a starlike node

5.1. The flow around an external corner

In the previous paragraph we have discussed the existence of higher-order conical stagnation points and their bifurcation due to analytical perturbations. In contrast to analytical perturbations also nonanalytical bifurcations may be encountered in conical flow patterns, especially in flow domains near discontinuities in the slope of the contour of a conical body. The existence of nonanalytical unfoldings may be shown by discussing the supersonic flow around an external corner.

An external corner configuration is composed of two plane delta wings Σ_1 and Σ_2 attached to each other along a common edge such that the planes of the wings include an angle with each other. The two remaining free (leading) edges are supersonic (velocity component normal to the edge is supersonic) thus the flows on either side of the configuration can be considered independently. The flow will be assumed to be conical and the centre of the conical field coincides with the apex of the configuration. We will be interested particularly in the flow in the region near the external angle, including both the case of the external axial corner, where the plane wings are nearly perpendicular to each other and the case where the configuration is similar to one side (upper or lower) of a delta wing with an arrow shaped cross-section (the plane $\bar{x} = 1$ in Fig. 2.18).

In order to describe the configuration in more detail, we use a right-handed cartesian coordinate system $\bar{x}, \bar{y}, \bar{z}$, with the origin in the apex and the positive $\bar{x}$-axis in the direction of the oncoming undisturbed flow. The $\bar{y}$-axis is chosen such that the leading edge of Σ_1 is in the $\bar{x}, \bar{y}$-plane and the $\bar{z}$-axis is perpendicular to it. The leading edge of wing Σ_2 and the $\bar{x}$-axis determine a plane Ω, which makes an angle ω with the $\bar{x}, \bar{y}$-plane measured positive as indicated in Figure 2.18. The leading edge of Σ_1 has a sweep angle Λ_1 with the $\bar{y}$-axis and Σ_1 is inclined with respect to the $\bar{x}, \bar{y}$-plane at an angle δ_1 (measured in the $\bar{z}, \bar{x}$-plane (see Fig. 2.18). The leading edge of Σ_2 has a sweep angle Λ_2 with the $\bar{y}, \bar{z}$-plane and Σ_2 makes an angle δ_2 with Ω (δ_2 is measured in a plane through the $\bar{x}$-axis and perpendicular to the plane Ω). The line where Σ_1 and Σ_2 meet will be called the corner line.

The conical symmetry of the flow field allows the flow to be described in the cross flow plane Σ normal to the corner line. The intersection of Σ_1 and Σ_2 with Σ are denoted by σ_1 and σ_2 respectively, they intersect in the corner point C. The obtuse angle between σ_1 and σ_2 is called the external angle and denoted by Φ_e; it can be determined from the parameters δ_1, δ_2, Λ_1, Λ_2 and ω.

The supersonic flow around an external corner and in particular the existence and character of conical stagnation points in this flow pattern has received much attention in the literature. In the symmetrical case (($\Lambda_1 = \Lambda_2$, $\delta_1 = \delta_2$) the corner point is a conical stagnation point; in the asymmetrical case the possibility that a conical Prandtl-Meyer fan will be formed at the corner point must be considered as well, see Salas and Daywitt (1979).

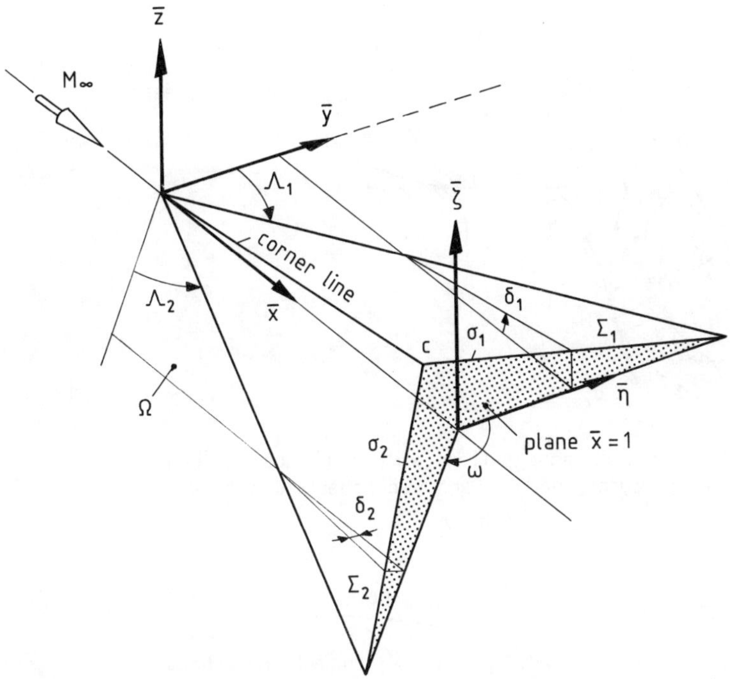

Fig. 2.18. Coordinate systems and geometry of the external axial corner.

We will restrict the discussion to the case that the corner point is a conical stagnation point. With reference to the local conical stagnation point solutions (paragraph 2 and 3), which have to satisfy the boundary conditions on σ_1 and σ_2, various possibilities for the type of conical stagnation points at the corner arise; such as an oblique saddle point or a starlike node. In numerical calculations, made by Kutler and Shankar (1976), both for symmetrical and asymmetrical configurations, an oblique saddle point was found at the corner point and a nodal point on each of the wing surfaces. Furthermore in these calculations the position of the nodal point corresponds to the point where the transverse pressure distribution on the wing surface attains a maximum. A qualitative impression of this flow pattern may be obtained from Fig. 2.19a.

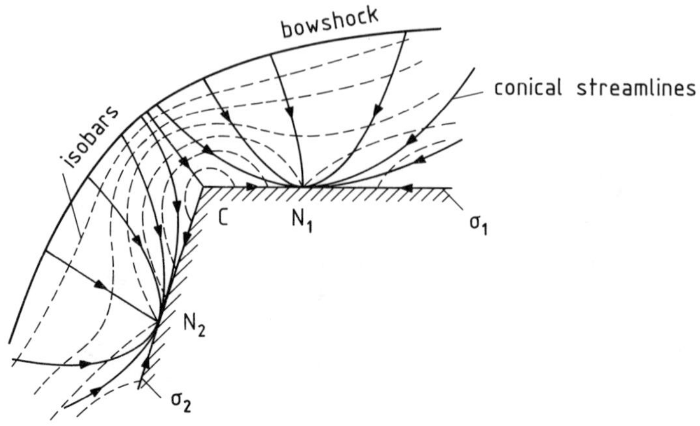

a) Conical streamline pattern with an oblique saddle point in the corner point (C) and two nodal points (N_1 and N_2)

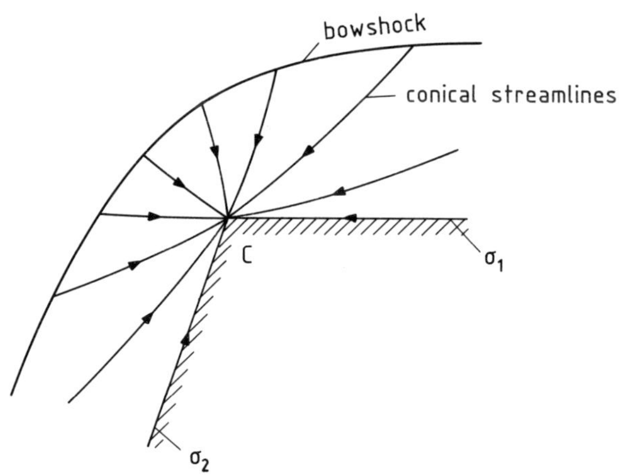

b) Conical streamline pattern with a single node at the corner point.

Fig. 2.19. Possible conical flow patterns near the corner point of an external corner.

In the symmetrical case this flow pattern was also found in the experiments of Bannink (1984). Despite the numerical evidence, Salas and Daywitt (1979) pointed out that a flow pattern as shown in Fig. 2.19b, having a nodal point at the cornerpoint as the only conical stagnation point in the flow field, should also be considered. In fact a numerical study made by Salas (1980) for symmetrical configurations ($\Lambda_1 = \Lambda_2 = 20°$, $\delta_1 = \delta_2 = 10°$) and a variation of ω from $\pi/2$ (external corner type) to π (delta wing type) seems to indicate the occurence of the flow pattern of Fig. 2.19b for ω close to π. Also, the calculations suggest a transition from the oblique saddle pattern to the starlike node pattern if ω is varied from $\pi/2$ to π.

In order to clarify this situation we want to investigate whether local conical stagnation point solutions applied to the region near the corner point support the idea of a single node pattern and the transition to the oblique saddle at the corner at some critical value of ω.

We note however, that the real nature of the conical stagnation point at the corner point can only be found by solving the full nonlinear boundary value problem without using any approximations such as numerical solutions. This, of course is a hard problem, unlikely to be solved. However, there is one noticeable exception, namely if $\delta_1 = \delta_2 = 0$, in which case Σ_1 and Σ_2 are aligned with the uniform flow. The resulting flow is the undisturbed uniform flow, which is represented by a single starlike node at the corner point. We will consider the local corner flow as a perturbation of this flow and the conical stagnation points at and near the corner point as bifurcations of this starlike node. The bifurcating solutions have to satisfy the boundary conditions on Σ_1 and Σ_2 and moreover in the limit $\delta_1, \delta_2 \to 0$ they have to revert to the uniform flow.

5.2. Boundary conditions and bifurcation modes

The analysis is performed in the cross-flow plane Σ, the reference system η, ζ has its origin in the corner point C and the η-axis coincides with σ_1.
If polar coordinates ρ,φ are used such that $\eta = \rho \cos \varphi$, $\zeta = \rho \sin \varphi$ then the corner boundaries σ_1 and σ_2 in Σ correspond to $\varphi = 0$ and $\varphi = \Phi_e$ respectively and the corner point C is given by $\rho = 0$.
Let a conical stagnation point exist at the corner point and let the conical potential F be expanded near the corner point by

$$F = F_o[1 + \rho^n F_n(\varphi) + \rho^m F_m(\varphi) + o(\rho^m)].$$

Then the local solutions:

$$F_n(\varphi) = \varepsilon_n \cos(n\varphi + \psi_n)$$

$$F_m(\varphi) = \beta_m \cos(m\varphi + \chi_m) \qquad , n < m < m_c$$

$$F_m(\varphi) = \beta_m \cos(m\varphi + \chi_m) + \gamma_m \cos(n\varphi + \psi_n) + \delta_m \qquad , m = m_c$$

obtained in paragraph 2 may be used, which have to satisfy the boundary conditions $\frac{\partial F}{\partial \varphi} = 0$ at $\varphi = 0$ and $\varphi = \Phi_e$.

With Eqs. (2.6) and (2.7) there is obtained

$$\psi_n = \chi_m = 0$$

$$n = k \frac{\pi}{\Phi_e} \quad , \quad k = 2,3,4,... \tag{2.35}$$

$$m = l \frac{\pi}{\Phi_e} \quad , \quad l = 3,4,5,...,$$

or if

$$m \neq l \frac{\pi}{\Phi_e}, \quad \text{for any } l = 3,4,5,... \text{ then } m = m_c, \ \beta_m = 0$$

Since external corner flows are considered for $\pi \leq \Phi_e \leq 2\pi$ and because $m > n > 1$ it follows that $l > k \geq 2$.

Obviously, imposing the boundary conditions on σ_1 and σ_2 is by itself insufficient to ensure the proper embedding in the surrounding main flow of the conical stagnation point solution near the corner. In fact, the freedom to further specify solutions is already expressed by the possibility of choosing the values of k and l in the leading terms of the expansion of the conical potential. The exponent n, occurring in the leading term is illustrated in Fig. 2.20 as a function of Φ_e for various values of k. The solutions for the various choices of k and l will be considered as perturbations on the uniform flow given by $F = F_o$ which has a starlike node at the corner point. Therefore, the coefficients in the expansion for F, such as ε_n, β_m etc. will be interpreted as small parameters tending to zero for the uniform flow. These small parameters vary in relation to each other such that $\rho^n F_n(\varphi)$ remains the leading term in the expansion, even if the parameters tend to zero.

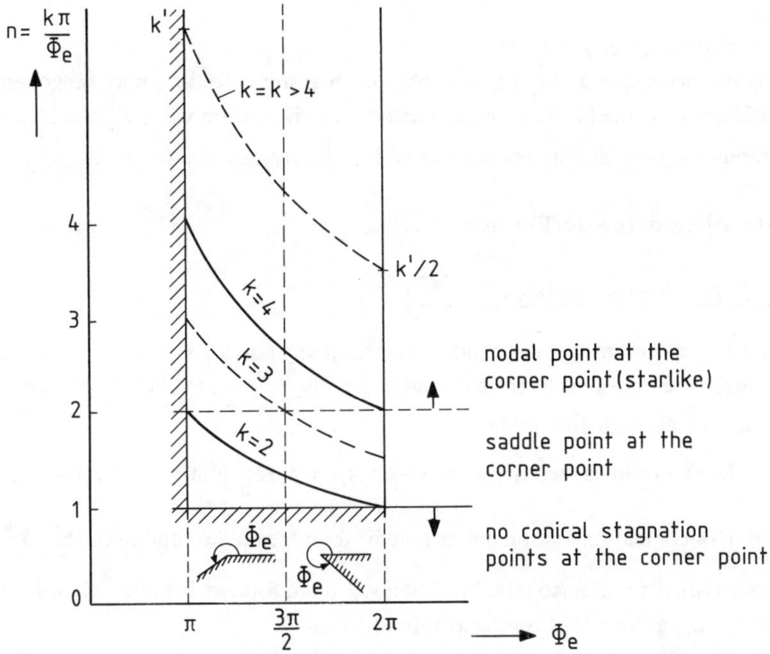

Fig. 2.20. Dependence of n on the external angle for different bifurcation modes (k = 2, 3, 4, ...).

As a result the flow near the corner point is obtained as an unfolding of the starlike node, the streamline pattern of which follows from the streamline equation in polar form by taking $F = F_o$:

$$\dot{\rho} = -F_o \cdot \rho$$

$$\rho\dot{\varphi} = 0$$

The unfolding of this system which satisfies the boundary conditions on σ_1 and σ_2, is

$$\dot{\rho} = F_o\{n\ \varepsilon_n\ \rho^{n-1}\ \cos(n\varphi) + O(\rho^{m-1}) - \rho\}$$
$$\rho\dot{\varphi} = F_o\{-n\ \varepsilon_n\ \rho^{n+1}\ \sin(n\varphi) + O(\rho^{m-1})\} \quad (2.36)$$

where n and m are given Eq. (2.35).
Since n is not necessary an integer it is obvious that the unfolding may be nonanalytic. The coefficient ε_n is the bifurcation parameter with bifurcation value $\varepsilon_n = 0$. Since k can take all integer values ≥ 2 it appears that various bifurcation modes arise.

5.3. Bifurcations of the starlike node

Case k = 2: first bifurcation mode [$n = \frac{2\pi}{\Phi_e}$]

In the first bifurcation mode, the exponent n can take values $1 < n \leq 2$ corresponding to external angles $\pi \leq \Phi_e < 2\pi$; the external angle $\Phi_e = 2\pi$ is excluded because the existence of a conical stagnation point requires $n > 1$.
From Eq. (2.35) it follows that $m = m_c = 3n\text{-}2$ for $1 < n \leq \frac{4}{3}$ and $m = 3n/2$ for $\frac{4}{3} \leq n \leq 2$.

The location of conical stagnation points may be found from the conditions that $\dot{\rho} = 0$ and $\dot{\varphi} = 0$ are satisfied simultaneously. For $1 < n < 2$ there follows from Eq. (2.36) for small values of ε_n, β_m as locations for these points:

$$C \quad : \rho = 0 \text{ (corner point)}$$
$$N_1, N_2 \quad : \varphi_1 = 0, \quad \varphi_2 = \Phi_e, \quad \rho_1 = \rho_2 = (n\ \varepsilon_n)^{\frac{1}{2-n}}\ ; \varepsilon_n > 0$$
$$N_3 \quad : \varphi_3 = \tfrac{1}{2}\Phi_e, \quad \rho_3 = (-n\ \varepsilon_n)^{\frac{1}{2-n}}\ ; \varepsilon_n < 0$$

where higher order terms in ε_n and β_m are omitted.
It appears that apart from the conical stagnation point at the corner (C), there are either two stagnation points N_1 and N_2 located on σ_1 and σ_2 respectively, or there is one point N_3, occurring as a free singularity in the flow field, Fig. 2.21a.
The points N_1 and N_2 have an infinite number of streamlines tangent to the body surface.

The nodal point N_3 is situated, to a first approximation, on the bisector of the external angle such that an infinite number of streamlines is tangent to this bisector.
Since $1 < n < 2$, the conical stagnation point at the corner is an oblique saddle point (this chapter, paragraph 3). The oblique saddle has three separatrices; two of them coincide with σ_1 and σ_2, the third with the bisector $\varphi = \frac{1}{2}\Phi_e$.

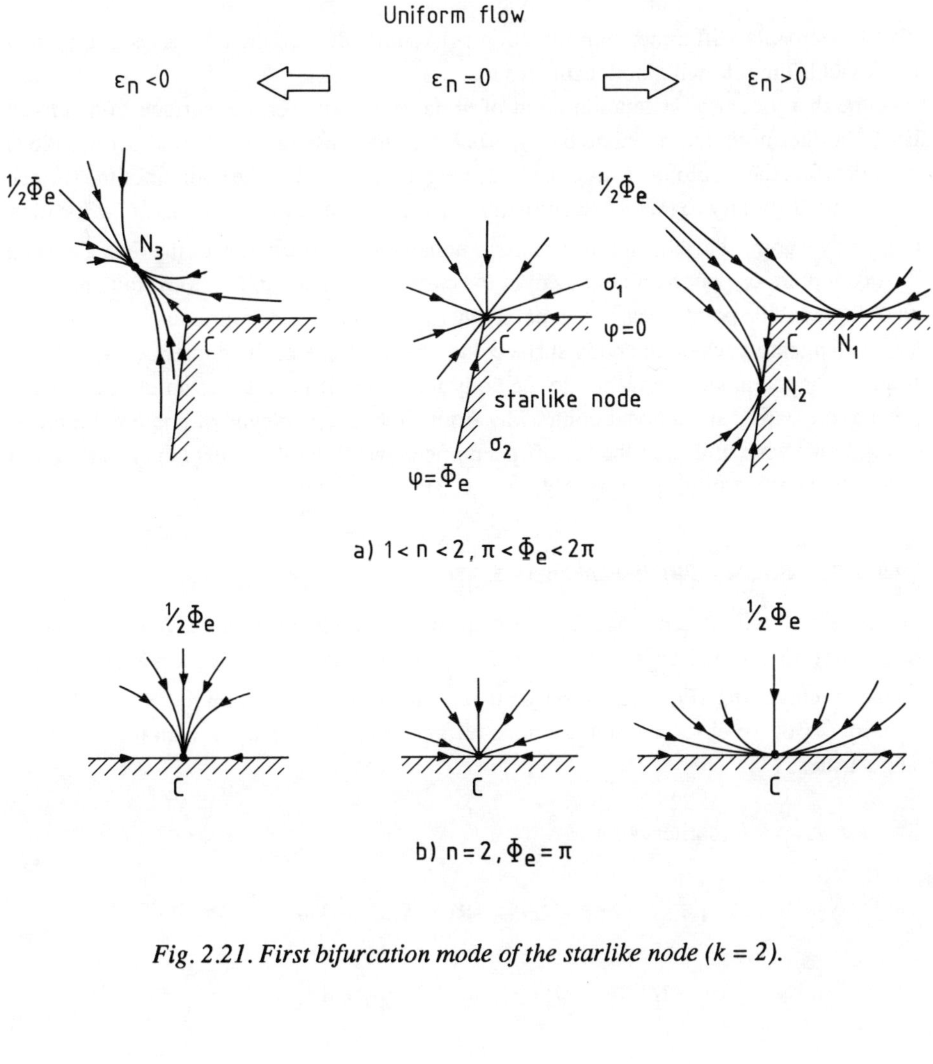

Fig. 2.21. First bifurcation mode of the starlike node (k = 2).

For n = 2, the external angle Φ_e equals π and the body contour at C shows no discontinuity in the slope. Therefore the classification of first-order singularities (paragraph 3) may be consulted to establish that for $\varepsilon_n \neq 0$ ($\varepsilon_n < \frac{1}{2}$) the 'corner' point C becomes a tangent node if $\varepsilon_n > 0$ and aan orthogonal node if $\varepsilon_n < 0$. Furthermore we observe that in this case there are no neighbouring conical stagnation points tending to C if $\varepsilon_n \to 0$, Fig. 2.21b.

This fundamental difference in bifurcation behaviour between the two cases $1<n<2$ and $n = 2$ yields the following noticeable result.
Assume that a conical stagnation point of nodal type exists on the surface of a smooth body. Furthermore, let this node be hyperbolic in the sense that analytical perturbations can not affect the topological structure in its neighbourhood. Let the contour of the surface be perturbed slightly so that a discontinuity in its slope is obtained ($\Phi_e > \pi$) at the location of the nodal point. This disturbance causes a nonanalytic perturbation of the flow field and the nodal point can not be maintained. In the case of a tangent node it must fall apart into an oblique saddle point (located at a point where the slope is discontinuous) accompanied by two tangent nodes on the body surface one on either side of the corner point.
If an orthogonal node is attached to the body the node bifurcates into an oblique saddle point at the corner and a nodal point away from the body. This type of nodal point bifurcation looks very similar to the lift-off phenomena which has been explained earlier as a saddle-node bifurcation (paragraph 4, section 3).

Case k = 3: second bifurcation mode $[n = \frac{3\pi}{\Phi_e}]$

In this second bifurcation mode (k = 3), the exponent n can take values $3/2 \leq n \leq 3$ corresponding to external angles $\pi \leq \Phi_e \leq 2\pi$. The possible range of n, together with the boundary conditions (Eq. (2.35)), shows that m satisfies the condition $m \geq 4\, n/3$. Consulting Eq. (2.36) we observe a starlike-node bifurcation for $3/2 \leq n < 2$ with the following singularities:

$$C \quad : \rho = 0 \text{ (corner point)}$$

$$N_1, N_3 \quad : \varphi_1 = 0, \quad \varphi_3 = 2/3\, \Phi_e, \quad \rho_1 = \rho_3 = (n\, \varepsilon_n)^{\frac{1}{2-n}} \quad ; \varepsilon_n > 0$$

$$N_2, N_4 \quad : \varphi_2 = 1/3\, \Phi_e, \quad \varphi_4 = \Phi_e, \quad \rho_2 = \rho_4 = (-n\, \varepsilon_n)^{\frac{1}{2-n}} \quad ; \varepsilon_n < 0$$

where higher order terms in ε_n and β_m are omitted.

The second bifurcation mode has in common with the first bifurcation mode that new conical stagnation points are generated from the original starlike node. The points N_1 and N_4 are located at the corner surfaces σ_1 and σ_2, respectively, whereas the points N_2 and N_3 appear as free singularities, Fig. 2.22a.
For $\varepsilon_n > 0$ only the conical stagnation points C, N_1 and N_3 exist and for $\varepsilon_n < 0$ the points C, N_2 and N_4 are present. It appears that all the points $N_{1,2,3,4}$ are stable nodes; N_1 and N_4 have an infinite number of streamlines tangent to the surface, N_2 and N_3 have an infinite number of streamlines tangent to the separatrices $\varphi = 1/3\ \Phi_e$ and $\varphi = 2/3\ \Phi_e$, respectively. The corner point C is an oblique saddle point with four separatrices; two of them coincide with $\sigma 1$ and σ_2 whereas the others approach C along $\varphi = 1/3\ \Phi_e$ and $\varphi = 2/3\ \Phi_e$. We conclude that in the second bifurcation mode for $3/2 \leq n < 2$ a starlike node bifurcates into an oblique saddle point at the corner point flanked by two nodes. One of these nodes lies on the corner surface (σ_1 or σ_2), whereas the second node appears as a free singularity in the flow.

<u>The case $n = 2$</u> corresponds to an external angle $\Phi_e = 3/2\ \pi$.
Since n is an integer the classification of first-order singularities (paragraph 3) may be consulted to obtain the flow pattern for various values of ε_n. For small values of ε_n a nodal point is formed at C with an infinite number of streamlines tangent to σ_1 for $\varepsilon_n > 0$ and tangent to σ_2 for $\varepsilon_n < 0$, Fig. 2.22b. In contrast to the case $3/2 \leq n < 2$ we observe that for $n = 2$ there are no conical stagnation points bifurcating from the starlike node.

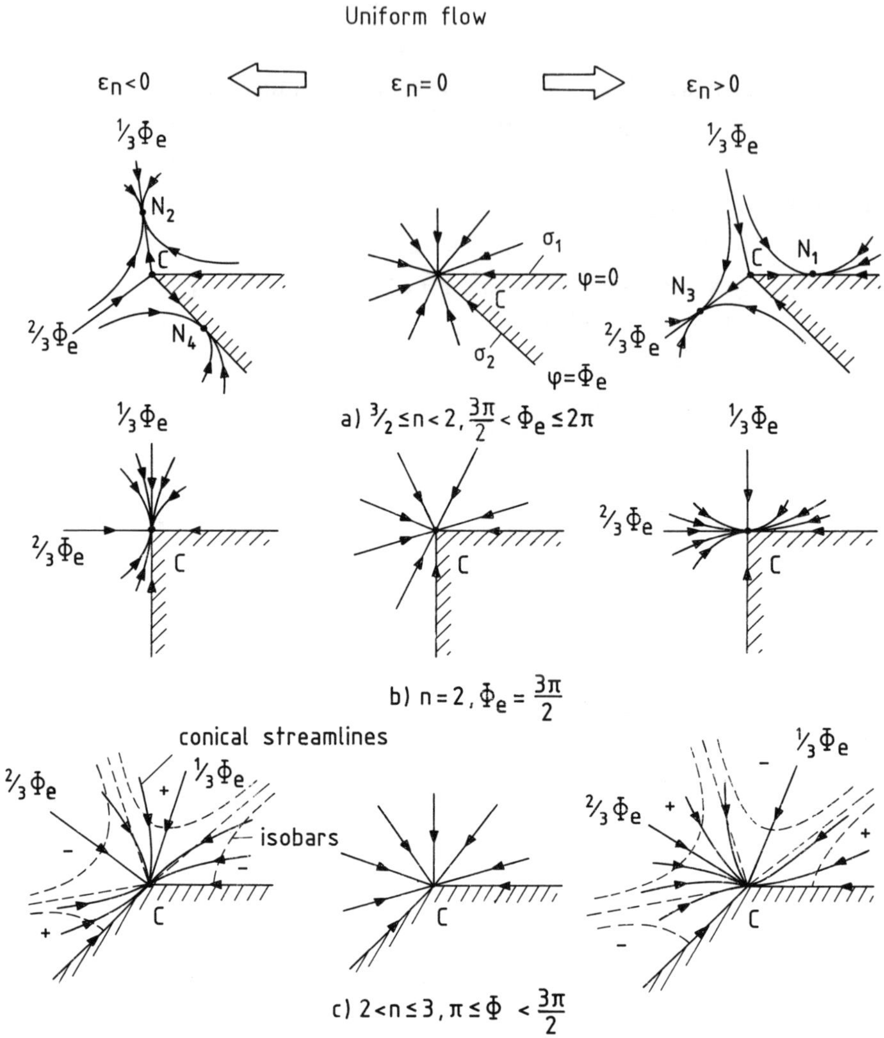

Fig. 2.22. Second bifurcation mode of the starlike node (k = 3).

In the case $2 < n \leq 3$ the external angle lies in the range $\pi \leq \Phi_e < 3/2\,\pi$. From Eq. (2.36) it follows that there are no conical stagnation points tending towards the corner point C for $\varepsilon_n \to 0$. For $\varepsilon_n \neq 0$ the corner point remains a starlike node. The rays $\varphi = 1/3\,\Phi_e$ and $2/3\,\Phi_e$ are to a first approximation conical streamlines which divide the flow field into three sectors. The conical streamlines in each sector are curved at the singular point, but

the sign of the curvature is opposite in adjacent sectors.
The corresponding flow patterns are sketched in Fig. 2.22c.
This figure gives also an impression of the pressure distribution near the corner which may be obtained from Eq. (2.8) and is given by

$$\left[\frac{p}{p_o}\right]^{\frac{\gamma-1}{\gamma}} = 1 - \frac{F_o^2}{1-F_o^2}\{-2(n-1)\,\varepsilon_n\,\rho^n \cos(n\varphi) + n^2\,\varepsilon_n^2\,\rho^{2n-2} + O(\rho^{\frac{7n}{3}-2})\}$$

The isobar pattern shows a saddle behaviour with separatrices at $\varphi = \frac{1}{6}\Phi_e, \frac{3}{6}\Phi_e$ and $\frac{5}{6}\Phi_e$ on which $p = p_o$.
It may be noted that in contrast to the first bifurcation mode, the second bifurcation mode is not symmetric with respect to $\varphi = \frac{1}{2}\Phi_e$.

<u>*Case k = 4: third bifurcation mode* $[n = \frac{4\pi}{\Phi_e}]$</u>

In the third bifurcation mode k = 4, the exponent n can take the values $2 \leq n \leq 4$ corresponding to external angles $\pi \leq \Phi_e \leq 2\pi$. This possible range of n, together with the boundary conditions, Eq. (2.35), shows that m satisfies the inequality $m > 5n/4$.
Then Eq. (2.36) reveals that, apart from the conical stagnation point at the corner point C, there are no neighbouring conical stagnation points in the flow field which tend to C for $\varepsilon_n \to 0$.
<u>For $n = 2$</u>, $\Phi_e = 2\pi$ the classification of first-order singularities may be applied and a streamline pattern with a single node at the corner point results, Fig. 2.23a.
<u>For $n > 2$</u> at the corner point there is a starlike node similar to that found in the second bifurcation mode. However, the number of sectors is now four and the rays $\varphi = 1/4\,\Phi_e$, $2/4\,\Phi_e$ and $3/4\,\Phi_e$ are the conical streamlines bordering these sectors. The corresponding flow patterns that occur in this bifurcation mode are sketched in Fig. 2.23b.
The pressure distribution near the corner, which may be obtained from Eq. (2.8) is given by

$$\left[\frac{p}{p_o}\right]^{\frac{\gamma-1}{\gamma}} = 1 - \frac{F_o^2}{1-F_o^2}\{-2(n-1)\,\varepsilon_n\,\rho^n \cos(n\varphi) + n^2\,\varepsilon_n^2\,\rho^{2n-2} + O(\rho^{\frac{9n}{4}-2})\}$$

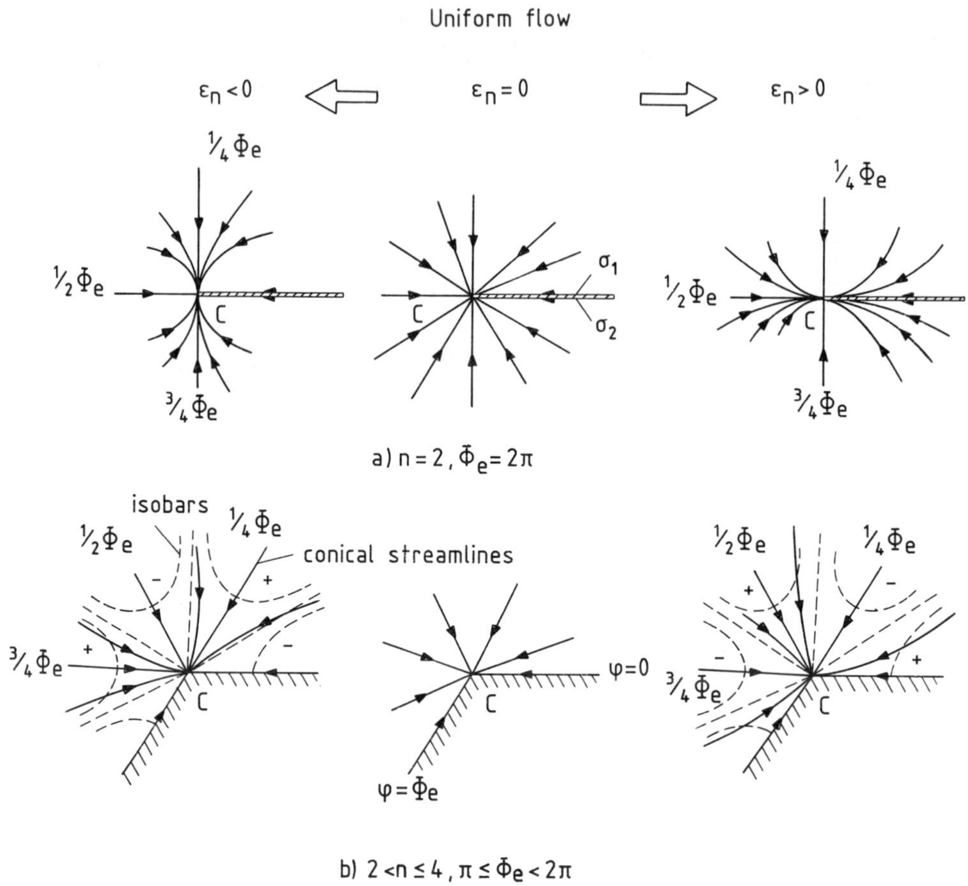

Fig. 2.23. Third bifurcation mode of the starlike node (k = 4).

The isobar pattern shows a saddle point behaviour with separatrices at $\varphi = 1/8\Phi_e$, $3/8\ \Phi_e$, $5/8\ \Phi_e$ and $7/8\ \Phi_e$ on which $p = p_o$. We note in particular that the pressure on the corner surface (σ_1 and σ_2) and on the bisector $\varphi = 1/2\ \Phi_e$ increases with the distance to the corner point for $\varepsilon_n > 0$ and decreases for $\varepsilon_n < 0$.

Case $k > 5$: higher bifurcation modes [$n = k\ \frac{\pi}{\Phi_e}$]

For $k > 5$ the exponent n satisfies $n > 2$ and $\pi \leq \Phi_e \leq 2\pi$. For the conical streamline pattern again the leading terms in Eq. (2.36) may be used, leading to a starlike node at the corner point, having k sectors. It may be noted that as $k \to \infty$ the flow pattern resembles more and more that of the uniform flow.

5.4. Symmetrical external corners

In order to illustrate the use of the classification of bifurcation modes of the starlike node, the flow around a symmetrical external corner will now be discussed. The symmetry implies that k = even, so that the generation of conical stagnation points from the starlike node only occurs in the first bifurcation mode (k = 2). In order to gain more insight into the question whether new conical stagnation points actually arise or not we will investigate the first mode in more detail; in particular we direct our attention to the question how the local corner flow fits within the overall flow field. The pressure distribution on the body surface and the location of the conical stagnation points as function of freestream Mach number and body geometry will receive special attention.

In this particular case the external angle Φ_e is related to wedge angle δ, sweep angle Λ, and ω by

$$\Phi_e = 2\pi - \cos^{-1}\left\{ \frac{2 \tan\delta \tan\Lambda \sin\omega + (1 - \tan^2\delta \tan^2\Lambda) \cos\omega - \tan^2\delta}{1 + \tan^2\delta + \tan^2\delta \tan^2\Lambda} \right\}$$

where $0 \le \cos^{-1} \le \pi$, since $\pi \le \Phi_e \le 2\pi$.
For the first bifurcation mode the expansion for the conical potential F may be written as (Eqs. (2.6), (2.7))

$$F = F_o \left[1 + \varepsilon_n \rho^n \cos(n\varphi) + \gamma_4\, \rho^{3n-2} \cos(n\varphi) + o(\rho^{3n-2})\right] \qquad (2.37)$$

where $\gamma_4 = \dfrac{(n\,\varepsilon_n)^3\, M_o^2}{4(2n-1)}$, $1 < n \le 2$, $\Phi_e = \dfrac{2\pi}{n}$.

The pressure distribution may be obtained using Eq. (2.8)

$$\left[\frac{p}{p_o}\right]^{\frac{\gamma-1}{\gamma}} = 1 - \frac{\gamma-1}{2} M_o^2 \left[n^2 \varepsilon_n^2 \rho^{2n-2} - 2(n-1)\varepsilon_n \rho^n \cos(n\varphi) + \right.$$

$$\left. 2n\, \gamma_4\, \varepsilon_n\{n + 2(n-1) \cos^2(n\varphi)\}\, \rho^{4n-4} + o(\rho^{4n-4})\right]$$

where p_o and M_o refer to values in the corner point.

Since we are particularly interested in the flow around an external corner with compressive surfaces ($\delta > 0$) (and it will be shown later that embedding of the local corner flow is only possible for $\varepsilon_n > 0$), we will restrict ourselves here to $\varepsilon_n > 0$ in which case a saddle point occurs at the corner and two nodal points on the body surfce at $\rho = \rho_N = (n\,\varepsilon_n)^{\frac{1}{2-n}}$

If the pressure distribution is written in terms of ρ_N, we obtain

$$\left[\frac{p}{p_o}\right]^{\frac{\gamma-1}{\gamma}} = 1 - \frac{\gamma-1}{2}\, M_o^2\, [\rho_N^{2\,(2-n)}\, \rho^{2n-2} - \frac{2(n-1)}{n}\, \rho_N^{2-n}\, \rho^n \cos(n\varphi)] + o(\rho^n) \tag{2.38}$$

At the corner point the isobars form a center point and on the body surface saddle points occur. These saddle points correspond to a minimum in the wall pressure, which, to first order, is given by

$$\frac{p_N}{p_o} = 1 - \frac{\gamma}{2}\, \frac{\Phi_e - \pi}{\Phi_e}\, M_o^2\, \rho_N^2$$

Matching of local corner flow with two-dimensional wedge flows

To the order indicated in Eq. (2.37) the expression for the conical potential F contains two free parameters ε_n and M_o, which may be used to match the local corner flow with the two-dimensional flow found in the region downstream of the supersonic leading edge. The flow in this region is similar to the flow over a wedge and will therefore further be referred to as two-dimensional wedge flow. The matching will be performed by requiring continuity of the velocity (in direction and magnitude) on the ray $\rho = \tilde{\rho}$, the intersection with the body surface of the Mach cone of the two-dimensional wedge flow emanating from the apex of the configuration. We then obtain

$$\begin{gathered} F_\rho = (F - \rho F_\rho)\,.\,\tan\tilde{\theta} \\ \left[1 + \frac{\gamma-1}{2} M_{2D}^2\right]\ \{(1+\rho^2)\, F_\rho^2 - 2\rho F F_\rho + F^2\} = \frac{\gamma-1}{2}\, M_{2D}^2 \end{gathered} \tag{2.39}$$

at $\varphi = 0$, Φ_e and $\rho = \tilde{\rho}$.

Here $\tilde{\theta}$ is the angle between the direction of the two-dimensional wedge flow and the corner line, and M_{2D} is the Mach number of the two-dimensional wedge flow. Substitution of Eq. (2.37) into Eq. (2.39) yields

$$n\varepsilon_n \tilde{\rho}^{n-1} + (3n-2)\gamma_4 \tilde{\rho}^{3n-3} = [1-(n-1)\varepsilon_n \tilde{\rho}^{n} - 3(n-1)\gamma_4 \tilde{\rho}^{3n-2} + o(\tilde{\rho}^{2n})] \,.\, \tan\tilde{\theta} \tag{2.40}$$

$$M_o^2[1+[1+\frac{\gamma-1}{2} M_{2D}^2]\{n^2\varepsilon_n^2 \tilde{\rho}^{2n-2} - 2(n-1)\varepsilon_n\tilde{\rho}^{n} + 2n(3n-2)\gamma_4\varepsilon_n\tilde{\rho}^{4n-4} + o(\tilde{\rho}^{4n-4})\}] = M_{2D}^2 \tag{2.41}$$

which gives ε_n, M_o; as a result also the location of the conical stagnation points and the pressure distribution may be deduced. It may be seen from Eq. (2.40) that, for small ε_n, ε_n and $\tilde{\theta}$ have the same sign which implies that for compressive wedge angles ($\delta > 0$, $\tilde{\theta} > 0$), $\varepsilon_n > 0$ and for expanding wedge angles ($\delta < 0$, $\tilde{\theta} < 0$), $\varepsilon_n < 0$.

For small values of $\tilde{\theta}$ the equations (2.40) and (2.41) appear to be uncoupled so that the unknown parameter ε_n can be obtained as a function of $\tilde{\theta}$ and $\tilde{\rho}$ from (2.40), then ε_n is approximated by

$$\varepsilon_n = \frac{\tan\tilde{\theta}}{n\tilde{\rho}^{\,n-1} + (n-1)\,\tilde{\rho}^{\,n}\,\tan\tilde{\theta}}$$

which gives the following approximation for the location of the nodal points on the wing surface σ_1 or σ_2 in the first bifurcation mode

$$\rho_N = \left(\frac{\tan\tilde{\theta}}{1+\frac{n-1}{n}\,.\,\tilde{\rho}\,\tan\tilde{\theta}}\right)^{\frac{1}{2-n}} \tilde{\rho}^{\,\frac{n-1}{2-n}} \tag{2.42}$$

Since $\tilde{\rho}$ can be expressed in terms of $\tilde{\theta}$ and M_{2D} by

$$\tilde{\rho} = \frac{\sqrt{M_{2D}^2 - 1}\,\tan\tilde{\theta} + 1}{\sqrt{M_{2D}^2 - 1} - \tan\tilde{\theta}}$$

the shift of the nodal point away from the corner point can be determined in terms of the physical variables Φ_e, $\tilde{\theta}$, M_{2D}. Figure 2.24 shows for three different Mach numbers of the two-dimensional wedge flow (M_{2D} = 1.5, 3 and 5) the location of the nodal point as a function of the external angle, Φ_e and the flow direction $\tilde{\theta}$. It may be observed that for external angles close to π (representating configurations of delta wing type) the nodal point lies very close to the corner point. This shift away from the corner becomes more apparent for increasing Mach numbers and a diverging of the flow towards the leading edge.

If the matching procedure is applied in the case of a symmetrical corner with compressive wedge angles, characterized by $\omega = \pi/2$, $\Lambda = 0$, the location of the conical stagnation points on the surfaces is found to depend on the parameters δ and M_∞ as shown in Fig. 2.25. Note that the results are given in the coordinates $\bar{\eta}\,[=\frac{\bar{y}}{\bar{x}}]$, $\bar{\zeta}\,[=\frac{\bar{z}}{\bar{x}}]$ which describe the cross flow plane $\bar{\Sigma}$ perpendicular to the $\bar{x}$ - axis. Remark that in the limit $\delta \rightarrow 0$ the nodal point shifts towards the corner point if compared with the position of the node that would appear for the flow around a single wedge, since then $(\bar{\eta})_N = 0$.

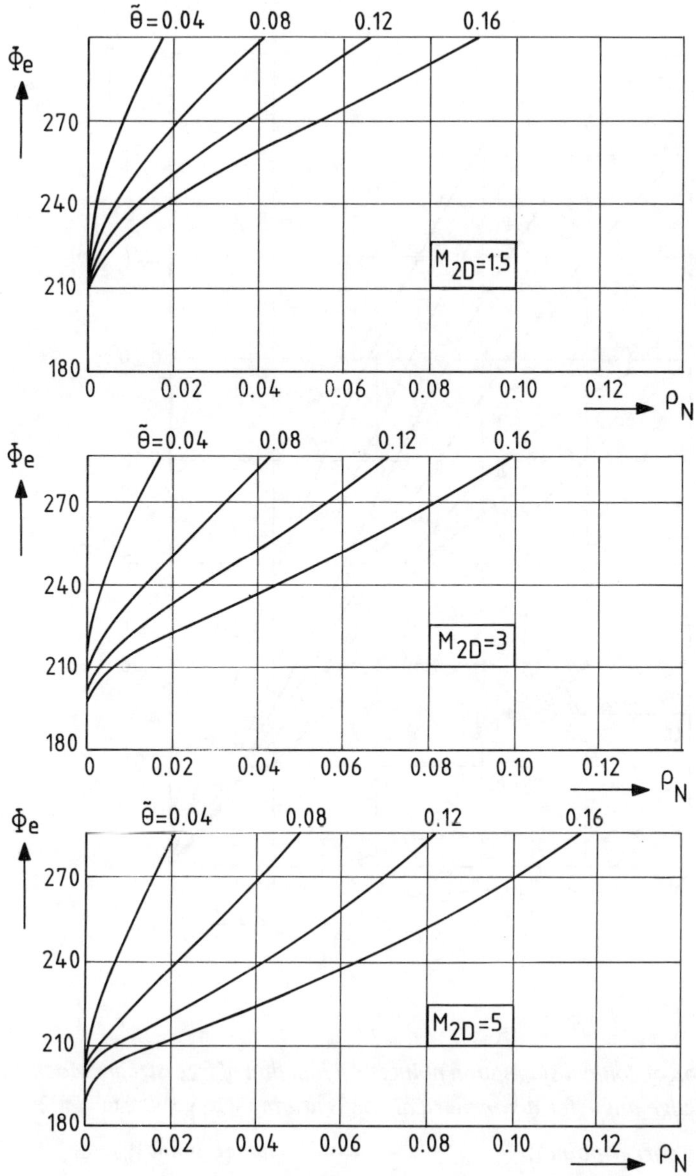

Fig. 2.24. Location of conical stagnation point on body surface as a function of flow direction $\tilde{\theta}$ and Mach number (M_{2D}) of the two-dimensional wedge flow.

Figure 2.25 seems to indicate that the bifurcation of the starlike node into an oblique saddle point and two nodal points is a higher order effect in δ, since for any finite Mach number M_∞ the lines for $\bar{\eta} = \bar{\eta}_C$ and $\bar{\eta} = \bar{\eta}_N$ seem to approach the origin at equal slope

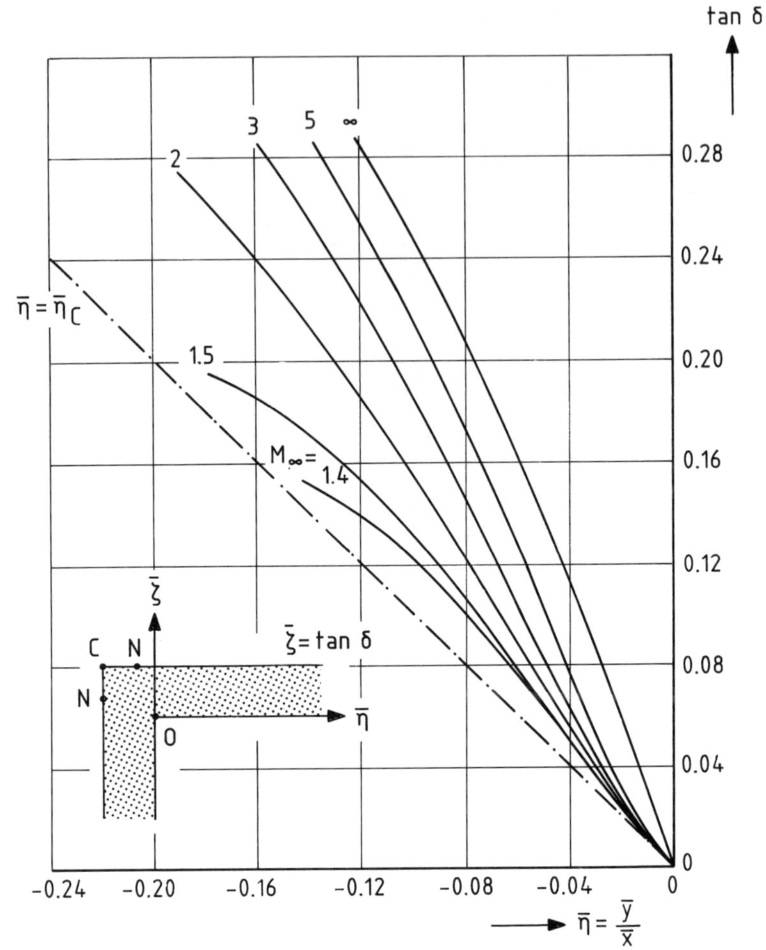

Fig. 2.25. Position of conical stagnation points as a function of free-stream Mach number and wedge angle for a symmetrical configuration wit $\Lambda = 0°$, $\omega = \pi/2$

—·—·— *corner point* $\bar{\eta}_C$ ——— *nodal point* $\bar{\eta}_N$

This effect may be verified by expanding Eq. (2.42) near $\delta = 0$ and taking into account that for $\omega = \pi/2$, $\Lambda=0$, $\tan\tilde{\theta} = \sin\delta$; then the leading term for ρ_N becomes

$$\rho_N = \tilde{\rho}^{\frac{1-n}{2-n}} \delta^{\frac{1}{2-n}}$$

Since on the wedge surface σ_1 we have $\tilde{\rho} = (\rho - \sin\delta)(1 + \rho\sin\delta)^{-1}(\cos\delta)^{-1}$ we may write:

$$\bar{\eta}_N = -\delta + \tilde{\rho}^{\frac{1-n}{2-n}} \cdot \delta^{\frac{1}{2-n}} + O(\delta^2),\ 1 < n < 2$$

$$\bar{\eta}_C = -\delta + O(\delta^3)$$

This calculation indicates that in a usual perturbation theory where δ is the perturbation parameter, such as used by Vorob'ev and Fedosov (1972), bifurcation of the starlike node is unlikely to appear.

In order to compare the pressure distribution on the wedge surfaces with numerical and experimental results we consider the case of $\delta = 10°$, $M_\infty = 3$; then $M_{2D} = 2.580$ and $\rho_N = 0.0783$. Applying Eq. (2.38) we obtain p/p_N which is shown in Fig. 2.26.
The agreement with numerical calculations of Salas (1980) and experimental results of Bakker et. al (1981) is quite satisfactory even at greater distance from the corner point. We note that in the numerical calculations entropy gradients are taken into account, whereas the present theory assumes potential flow near the corner, and only a limited number of terms in the expansion are used.

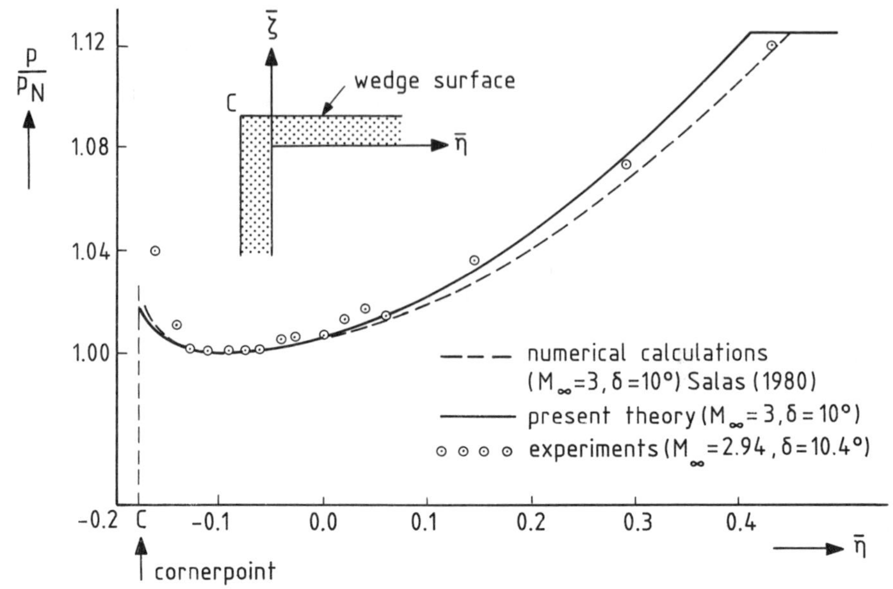

Fig. 2.26. Pressure distribution along the wedge surface of symmetrical external corner with $\Lambda = 0°$, $\omega = \pi/2$.

5.5. Transition of oblique saddle to starlike node

Theoretical observations

So far the use of the first bifurcation mode seems a correct way to describe the flow around a symmetrical external corner with compressive surfaces. As a result two nodal points are found on the body surface as well as a saddle point at the corner. This conclusion is unaffected by a change of ω from $\omega \approx \pi/2$ (the corner type configuration) to $\omega \approx \pi$ (the delta wing type). In fact numerical calculations were performed by Salas for symmetrical configurations ($\delta = 10°$, $\Lambda = 20°$ and $40°$), $M_\infty = 3$ and values of ω ranging from 90° to 180°.

According to Salas, these calculations indicate that for higher values of ω nodal points on the surface are not present and that the corner point is then a nodal point instead of a saddle point.

This would imply that for a certain value of ω a transition from the first bifurcation mode (k = 2) to a higher bifurcation mode (k = 4,6,...) would take place. However such a transition also implies the sudden disappearance of the nodal point situated away from the corner. This is impossible since the nodal point away from the corner is structurally stable

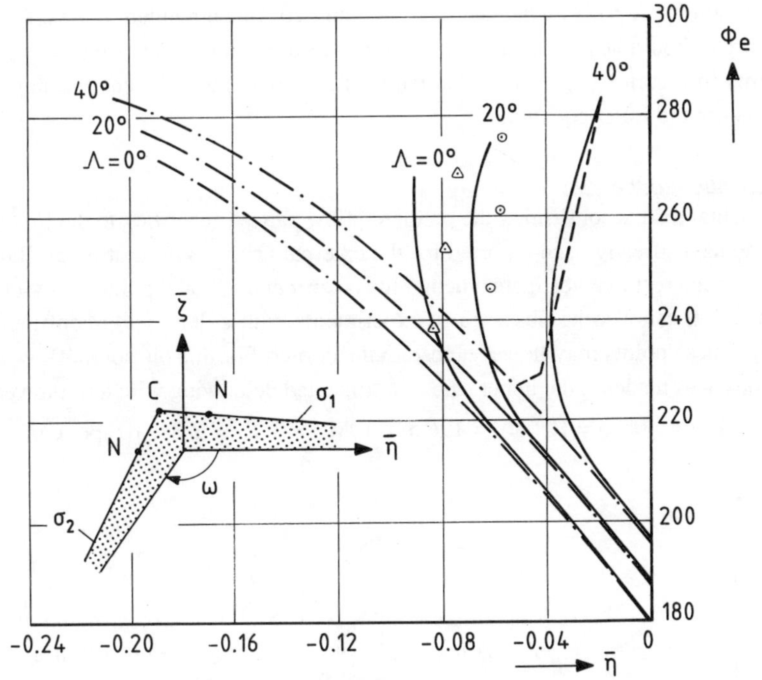

Fig. 2.27. The conical stagnation point on σ_1,
$M_\infty = 3$, $\delta = 10$.
—·—·— *corner* ——— *node*
- - - - - - *Salas (1980);*
△ ⊙ *Salas (priv. comm.)*

with respect to changes in the flow due to variation of Φ_e. We further remark that if the nodal points are close to the corner point, they are very difficult to detect in numerical calculations, in contrast to the situation when they can be clearly observed further away from the corner point.

In order to show how close the nodal points are to the corner point as ω tends to 180° for the configuration $M_\infty = 3$, $\delta = 10°$ at $\Lambda = 0°$, 20° and 40°, we calculated the position of the conical stagnation points according to the first bifurcation mode; as shown in Fig. 2.27. For these calculations it appears that for $\Phi_e < \sim 220°$ the nodal points and the corner point are so close that a numerical detection would be rather unlikely. For $\Phi_e > \sim 220°$ there is a substantial deviation of the nodal points away from the corner point. If we compare the results of the present theory with the numerical results of Salas (1980) there is a good agreement for $\Lambda = 40°$, whereas for $\Lambda = 0°$ and $20°$ the agreement is less. As mentioned before it should be noted that the present results are obtained by using a finite number of terms in a series expansion of potential flow solutions, whereas the numerical calculations use the Euler equations.

<u>*An experimental observation*</u>

As experimental results confirming the predicted flow pattern according to the first bifurcation mode have already been reported for the external angle configuration by Bannink (1984), it is of interest to investigate whether the occurrence of nodal points, distinct from the corner point, can also be observed in experiments with a delta wing configuration, even though these points may lie very close to the corner. For this purpose a flow visualization study was made on the upper side of a truncated delta wing with a flat lower surface ($\omega=\pi$ rad, $\Lambda = 40°$, $\delta = 10°$, $\Phi_e = 196.5°$) at $M_\infty = 3$, $Re = 2.3 \times 10^5$ per cm.

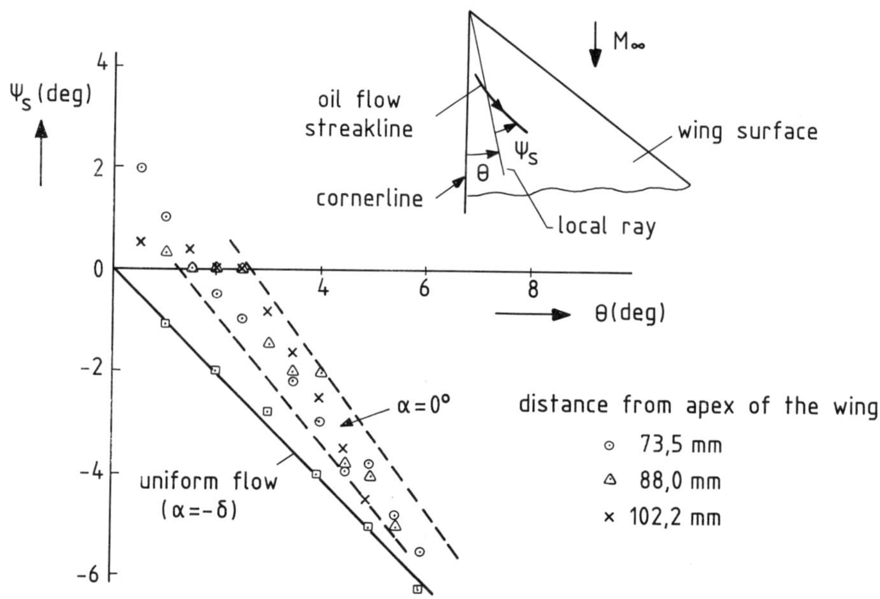

Fig. 2.28. Direction of oil flow streaklines.

Although one should be cautious in drawing conclusions on inviscid flow patterns from flow visualization techniques, experiments on external corners performed by Salas & Daywitt (1979) and Bannink (1984) have shown that the conical nature of the flow is borne out by the oil flow streaklines. The oil streaklines on the upper face of the wing are photographed when the incidence α of the plane lower surface (measured in the symmetry plane) is set at 0°.
From this picture the angle ψ_s of the local streamline with the local ray has been carefully measured. The results are collected in Fig. 2.28; they do suggest that nodal points may be distinguished in the conical flow field. The reliability of these results may be judged by comparing them with those obtained for $\alpha = -\delta = -10°$, in which case the upper surface is aligned with the flow. Comparing the results for $\alpha = 0$ and $\alpha = -10°$, a certain shift of the nodal singularity away from the corner is observed.

6. References

Bakker, P.G. and Bannink, W.J. (1974)
Conical stagnation points in the supersonic flow around slender circular cones at incidence. Delft University of Technology, Report VTH-184.

Bakker, P.G. (1977)
Conical streamlines and pressure distribution in the vicinity of conical stagnation points in isentropic flow. Delft University of Technology, Report LR-244.

Bakker, P.G., Bannink, W.J. and Reyn, J.W. (1981)
Potential flow near conical stagnation points, J. Fluid Mech., Vol. 105.

Bakker, P.G. (1984)
Structural stability and bifurcation in conical flow fields. Delft University of Technology, Report LR-424.

Bakker, P.G. and Reyn, J.W. (1985)
Conical flow near external axial corners as a bifurcation problem. AIAA-Journal, Vol. 23, No. 1.

Bannink, W.J. and Nebbeling C. (1978)
Measurements of the supersonic flow field past a slender cone at high angles of attack. AGARD Conference Proceedings 247, Paper 22.

Bannink, W.J. (1984)
Investigation of the conical flow field about external axial corners. AIAA-Journal, Vol. 22, No. 3.

Bulakh, B.M. (1970)
Nonlinear conical flow. Translated from the Russian by J.W. Reyn and W.J. Bannink, Delft University Press (1985).

Busemann, A. (1929)
Drücke auf kegelförmige Spitzen bei Bewegung mit Überschallgeschwindigkeit. ZAMM, 9 (6)

Ferri, A. (1951)
Supersonic flow around circular cones at angles of attack. NACA TR-1045.

Fowell, L.R. (1956)
Exact and approximate solutions for the supersonic delta wing. J. Aeron. Sci. 23 (8).

Germain, P. (1949)
La théorie génerale des mouvements coniques et ses applications à l'aérodynamique supersonique. ONERA, Publ. no. 34.

Goldstein, S. & Ward, G.N. (1950)
The linearized theory of conical fields in supersonic flow with application to plane airfoils. Aeron. Quart. II.

Grossman, B. (1979)
Numerical procedure for the computation of irrotational conical flows. AIAA-Journal, Vol. 17, No. 8.

Kutler, P. and Shankar, V. (1976)
Computation of the inviscid supersonic flow over external axial corners. Proceedings of the Heat Transfer and Fluid Mechanics Institute, Davis. Calif.

Maslen, S. (1952)
Supersonic conical flow. NACA TN-2651.

Melnik, R.E. (1967)
Vortical singularities in conical flow. AIAA-Journal, Vol. 5, No. 4.

Reyn, J.W. (1960)
Differential geometric considerations on the hodograph transformation for irrotational conical flow. Archive Rat. Mech. Anal. 6(4).

Salas, M.D. and Daywitt, J. (1979)
Structure of the conical flow field about external axial corners. AIAA-Journal, Vol. 17, No. 1.

Salas, M.D. (1980)
Careful numerical study of flow fields about symmetrical external conical corners. AIAA-Journal, Vol. 18, No. 6.

Smith, J.H.B. (1972)
Remarks on the structure of conical flow. Progress in Aerospace Sciences, Vol. 12, Pergamon.

Stocker, P.M. and Mauger, F.E. (1962)
Supersonic flow past cones of general cross-section.
J. Fluid Mechanics, Vol. 13 (3).

Taylor, G.I. and Macoll, J.W. (1933)
The air pressure on a cone moving at high speeds. Proc. Roy. Soc., Ser. A 139.

Vorob'ev, N.F. and Fedosov, V.P. (1972)
Supersonic flow around a dihedral angle (conical case). Izvestiya Akademii Nauk SSR, Mekhanika Zhidkostii i Gaza, No. 5.

CHAPTER III. TOPOLOGICAL ASPECTS OF STEADY VISCOUS FLOWS NEAR PLANE WALLS

1. A way to obtain local solutions of the Navier-Stokes equations

In this chapter the qualitative theory of dynamical systems will be applied to steady two-dimensional viscous flows of continua near fixed and moving boundaries. The topological properties of these flows may be derived from solutions of the governing differential equations. Qualitative considerations about these solutions in the phase plane will be presented with the aim to provide insight into the physics of the flow pattern prior to solving appropriate, initial/boundary value problems.

The topology of steady viscous flows is studied on the basis of *local* solutions of the Navier-Stokes equations (NS) governing incompressible laminar flow. These local solutions are obtained by performing Taylor expansions for the velocity vector field near a point in the flow field. The steady streamline pattern near this point is represented by the trajectories of the second-order system $\dot{x} = u(x,y)$, $\dot{y} = v(x,y)$ with u,v denoting the velocity components in a cartesian reference system x,y.
The Taylor expansion of the velocity field is not restricted to the neighbourhood of regular points ($u \neq 0$, $v \neq 0$) only but may also be performed near separation or attachment points ($u = 0$, $v = 0$) because, as Dean (1950) has shown, the NS-equations allow also for analytical solutions near points where u and v vanish simultaneously.

Consider the flow pattern that results if an analytic velocity vector field is expanded up to the N-th order near an arbitrary point O, then the trajectory pattern near O is governed by the system (S):

$$\dot{x} = u = \sum_{i=0}^{N} \sum_{j=0}^{N-i} U_{i,j} + O(N+1)$$

$$\dot{y} = v = \sum_{i=0}^{N} \sum_{j=0}^{N-i} V_{i,j}\, x^i y^j + O(N+1) \qquad \text{(S)}$$

where O(N+1) denote higher-order terms of at least order N+1, composed by powers of x and y. The coefficients $U_{i,j}$ and $V_{i,j}$ are treated as constants and are left unknown in a local analysis. Since the flow satisfies the continuity equation and the NS-equations, relationships between $U_{i,j}$ and $V_{i,j}$ exist, reducing the number of coefficients that can be chosen independently. A further reduction of unknowns is obtained by partially fulfilling boundary conditions; for example if one studies flow patterns near a wall where no-slip conditions hold.

Our main objective will be to give a unified description of all topologically different flow patterns that will arise near O if the remaining coefficients in (S)are varied independently. At first sight this task seems hard since system (S) contains an infinite number of terms. To overcome this difficulty we content ourselves to study phase portraits of a truncated system (S_N) which is derived from (S) by omitting the O(N+1) terms.

$$\dot{x} = u = \sum_{i=0}^{N} \sum_{j=0}^{N-i} U_{i,j}\, x^i y^j$$

$$\dot{y} = v = \sum_{i=0}^{N} \sum_{j=0}^{N-i} V_{i,j}\, x^i y^j \qquad (S_N)$$

Since (S_N) contains a finite number of terms the question arises whether (S_N) is appropriate to describe the local flow topology of real flows or more precisely, which conditions have to be fulfilled so that the systems (S) and (S_N) have equal toplogical structure near O. Can N be chosen such that it suffices to study the trajectory pattern of system (S_N) instead of that of the original system (S).

With regard to this question we know that important elements of phase portraits are the singular points, the local trajectory pattern near these points and the position of separatrices.

Thus if singular points (and their local trajectory pattern) of system (S_N) keep their qualitatively character even if higher order terms are added, then the truncated system (S_N) suffices to obtain a qualitative description of the local flow topology.

For instance near hyperbolic singularities only linear terms are of importance so that the truncated system is simply found by taking N = 1.

For isolated non-hyperbolic singularities also a truncated system can be found provided

that N is sufficiently large so that O(N+1) terms do not disturb the flow topologically. Center manifold theory and blow up techniques will be used to examine which higher order terms play a significant role.
This implies that for a given N a classification of degenerate singularities can be established and each of them represents a local flow topology which can occur in flows that are governed by the full NS-equations.

Apart from these relatively simple local flow patterns containing only one singularity, system (S_N) enables us also to encounter more complicated flow patterns consisting of a *cluster* of singular points forming a coherent streamline structure.
In order to ascertain that such a cluster is a local solution of the NS-equations, it will be sought as an *unfolding* of a nonhyperbolic or higher-order singularity of system (S_N). Let this statement be explained as follows.
As we know, higher-order or degenerate singularities of (S_N), being local solutions of the NS-equations appear under certain specified conditions involving some of the coefficients $U_{i,j}$ and $V_{i,j}$ to be zero. The unfolding of the degeneracy follows by inserting again the terms whose vanishing in (S_N) did cause the appearance of the degenerate singularity. To obtain the unfolding only lower-order terms have to added and thus the corresponding unfolding is also a local solution of the governing flow equations.
This important conclusion enables us to state that higher-order singularities of system (S_N), although they might seem not to be important in practice because of their rare appearance, serve as corner-stones to obtain local solutions of the NS-equations displaying flow patterns having topologies of increased complexity.

The method to study higher-order singularities and their unfoldings, as outlined so far, is worked out in the subsequent paragraphs. The structurally stable flow patterns, among those obtained as bifurcations of a degenerate singularity, are interpreted physically; some of them appear to be very familiar in aerodynamic applications, others are new and are concerned with topics on laminar flows like the genesis of laminar separation bubbles, flow separation near moving walls, interference effects in separation and reattachment flows, and formation of asymmetric standing eddies in the near wake behind a body.

2. Steady viscous flow near a plane wall, elementary singular points in the flow patterns

2.1 Approximate solutions near a plane wall

Consider the incompressible two-dimensional steady laminar flow in the neighbourhood of a plane wall at rest. The flow will be described in a cartesian coordinate system x,y such that the x-axis coincides with the wall surface; x is increasing in downstream direction. The y-axis is chosen perpendicular to the wall surface such that the flow near the wall occurs in the upper halfplane $y \geq 0$.
The velocity components in the x and y direction will be denoted by u and v respectively while the pressure is indicated by the symbol p.
The governing equations for steady flow are:

continuity equation:

$$\frac{\partial u}{\partial x} + \frac{\partial v}{\partial y} = 0 \tag{3.1}$$

NS-equations conserving momentum:

$$u \frac{\partial u}{\partial x} + v \frac{\partial u}{\partial y} + \frac{\partial p^*}{\partial x} = \nu \nabla^2 u$$

$$u \frac{\partial v}{\partial x} + v \frac{\partial v}{\partial y} + \frac{\partial p^*}{\partial y} = \nu \nabla^2 v \tag{3.2}$$

p^* denotes the kinematic pressure $\frac{p}{\rho}$, ν is the kinematic viscosity of the fluid and ∇^2 stands for the well-known Laplace operator.
The kinematic pressure can be eliminated from Eq. (3.2) in order to obtain a single equation for the vorticity $\omega = \frac{\partial v}{\partial x} - \frac{\partial u}{\partial y}$,

$$u \frac{\partial \omega}{\partial x} + v \frac{\partial \omega}{\partial y} = \nu \nabla^2 \omega \tag{3.3}$$

the so-called vorticity transport equation.

On the wall surface where $y = 0$ the flow satisfies the no-slip boundary conditions

$$u(x,o) = v(x,o) = 0$$

If the velocity components u and v are analytic functions of x and y and have to satisfy the continuity equation and the boundary conditions at $y = 0$, they can be written

$$u = y\,\hat{u}(x,y), \quad v = y^2\,\hat{v}(x,y)$$

where $\hat{u}(x,y)$ and $\hat{v}(x,y)$ are analytic functions. To describe the viscous flow near an abitrary point on the wall (say at $x = 0$) let u and v be expanded as

$$u = y(a_1+a_2x+a_3y+a_4x^2+a_5xy+a_6y^2+a_7x^3+a_8x^2y+a_9xy^2+a_{10}y^3) + O(5)$$

$$v = y^2(b_1+b_2x+b_3y+b_4x^2+b_5xy+b_6y^2) + O(5) \tag{3.4}$$

where O(5) denotes fifth- and higher-order terms, a_i and b_j being constant coefficients. Since the flow is supposed to be a solution of the steady Navier-Stokes equation, relationships between the coefficients a_i and b_j will appear.

The continuity equation will be satisfied (up to third-order terms) if the following conditions hold:

$$a_2 + 2b_1 = 0 \qquad a_4 + b_2 = 0 \qquad 3a_7 + 2b_4 = 0$$

$$a_5 + 3b_3 = 0 \qquad 2a_8 + 3b_5 = 0 \tag{3.5}$$

$$a_9 + 4b_6 = 0$$

The vorticity transport equation is satisfied (to the first-order) if

$$2a_4 + 3a_6 = 0$$

$$2a_7 + a_9 = 0 \tag{3.6}$$

$$8\nu\,(a_8 + 3a_{10}) = a_1\,a_2$$

Equations (3.4) together with Eqs. (3.5) and (3.6) form a consistent set of relations to describe the viscous flow near an arbitrary point at the wall with fourth-order accuracy. Note that in the given order estimation all coefficients b_j can be expressed in terms of the coefficients a_i.

The unknown coefficients occurring in Eq. 3.4 can be used for a proper fitting of the local flow into a surrounding main flow. The number of these coefficients that can be chosen independently is reduced considerably by the relations given by Eqs. (3.5) and (3.6). For a fourth-order expansion, as given in Eq. (3.4), only seven coefficients (a_1, a_2, a_3, a_4, a_5, a_7 and a_8) are left instead of the original sixteen.

Our main objective here will be a description of all topologically different flow patterns that will arise near (0,0) if the seven unknowns are varied independently. Moreover, the conditions for which these flow patterns appear will be formulated. To interpret these conditions in a physical context it is worthwhile to relate the coefficients a_i to wall shear stress, pressure and gradients of wall shear stress and pressure. The wall shear stress is usually defined as $\tau = \mu \left[\frac{\partial u}{\partial y}\right]_{y=0}$ where μ is the dynamic viscosity of the fluid. Consulting Eqs (3.4) and (3.5) one can verify the relations:

$$
\begin{aligned}
a_1 &= \frac{\tau}{\mu} \\
a_2 &= \frac{\tau_x}{\mu} && = -\frac{p_y}{\mu} \\
a_3 &= && = +\frac{1}{2}\frac{p_x}{\mu} \\
a_4 &= \frac{1}{2}\frac{\tau_{xx}}{\mu} && = -\frac{1}{2}\frac{p_{xy}}{\mu} \\
a_5 &= && = \frac{1}{2}\frac{p_{xx}}{\mu} && = -\frac{1}{2}\frac{p_{yy}}{\mu} \\
a_7 &= \frac{1}{6}\frac{\tau_{xxx}}{\mu} && = -\frac{1}{6}\frac{p_{xxy}}{\mu} && = \frac{1}{6}\frac{p_{yyy}}{\mu} \\
a_8 &= && = \frac{1}{4}\frac{p_{xxx}}{\mu} && = -\frac{1}{4}\frac{p_{xyy}}{\mu}
\end{aligned}
\qquad (3.7)
$$

which hold in the point (0,0).
Indices are used to indicate partial derivatives.

The streamline pattern near the wall in the upper half plane $y \geq 0$ may be derived from the velocity field $u(x,y)$, $v(x,y)$ as

$$\dot{x} = \frac{dx}{dt} = u(x,y)$$
$$\dot{y} = \frac{dy}{dt} = v(x,y) \tag{3.8}$$

t denotes a time-like parameter varying along streamlines.

Introducing the Taylor expansion of the velocity components from Eq. (3.4) into (3.8) we obtain a nonlinear autonomous system for the description of the streamline pattern; the phase plane represents the real flow domain, the phase trajectories may be identified with the streamlines and t serves as a parameter varying along a trajectory.
Since y is a common factor in $u(x,y)$ and $v(x,y)$, the line $y = 0$ is a line of singular points implying a quasi nonhyperbolic singular character at the wall surface. Since the trajectories of system (3.8) are identical with those of the equivalent system

$$\dot{x} = y^{-1}\, u(x,y), \quad \dot{y} = y^{-1}\, v(x,y) \tag{3.9}$$

with $(\dot{\ }) = \dfrac{d}{dt^*} = \dfrac{1}{y}\dfrac{d}{dt}$, we can remove the singular character of the line $y = 0$ by investigating Eq. (3.9) instead of Eq. (3.8). In Eq. (3.9), t* is a time-like parameter varying along the trajectories (streamlines).
The dynamical system under consideration becomes

$$\dot{x} = a_1+a_2x+a_3y+a_4x^2+a_5xy - \frac{2}{3}\, a_4y^2 + a_7x^3 + a_8x^2\, y - \frac{1}{2}\, a_7xy^2 + \frac{1}{3}\left(\frac{a_1a_2}{8\nu} - a_8\right) y^3 + O(4)$$
$$\dot{y} = -\frac{1}{2}\, a_2y - a_4xy - \frac{1}{3}\, a_5y^2 - \frac{3}{2}\, a_7x^2\, y - \frac{2}{3}\, a_8xy^2 + \frac{1}{2}\, a_7y^3 + O(4) \tag{3.10}$$

The elementary singular points of this system will be analysed below (section 2.2 and

2.3); the classification of higher-order singularities will follow in a separate paragraph (paragraph 3).

2.2 Elementary singular points located at the wall (on-wall singularities)

Let us analyse the flow near a point on the wall where system (3.10) displays a singularity. We take the origin of our reference system in the singular point, then $a_1 = 0$ which means that the wall shear stress vanishes in the singularity; $\tau(0,0) = 0$. System (3.10) may now be written:

$$\dot{x} = a_2x + a_3y + a_4x^2 + a_5xy - \frac{2}{3}a_4y^2 + O(3)$$

$$\dot{y} = -\frac{1}{2}a_2y - a_4xy - \frac{1}{3}a_5y^2 + O(3)$$

The eigenvalues of the linear part yield $\lambda_1 + \lambda_2 = \frac{1}{2}a_2$ and $\lambda_1\lambda_2 = -\frac{1}{2}a_2^2$, thus if $a_2 \neq 0$ the singular point at (0,0) is a (hyperbolic) *saddle* and the linear part suffices for the description of the flow topology. The saddle point has an oblique separatrix in the direction $y = \frac{-3a_2}{2a_3}x$ and a separatrix along the wall surface ($y=0$). The flow patterns that arise for various values of a_2 ($a_3>0$) are shown in Figure 3.1; for $a_2 < 0$ the flow separates from the wall and for $a_2 > 0$ flow attachment occurs. The flow pattern for $a_3 < 0$ may be obtained by the transformation $x \rightarrow -x$.

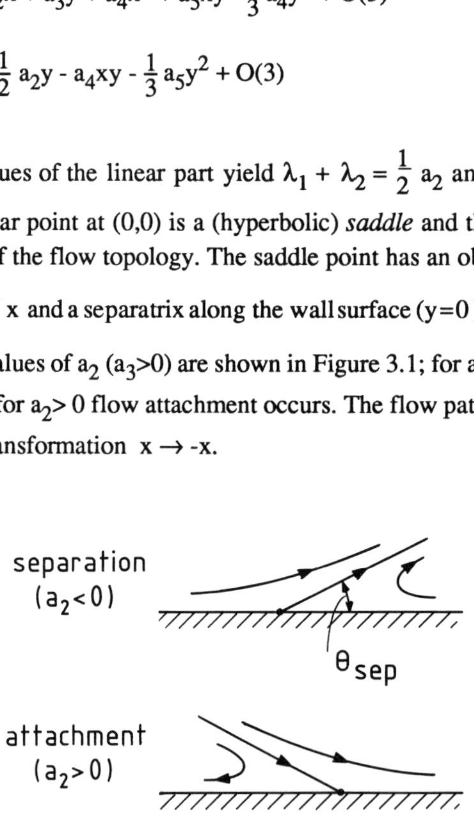

Fig. 3.1. Oswatitsch-Legendre solution, hyperbolic singularity on the wall.

With Eq. (3.7) the separation angle θ_{sep} may be written in terms of shear stress gradient and pressure gradient at the separation point as

$$\tan \theta_{sep} = -\frac{3}{2}\frac{a_2}{a_3} = -\frac{3\,\tau_x(0,0)}{p_x(0,0)} = \frac{3\,p_y(0,0)}{p_x(0,0)} \tag{3.11}$$

This classical result was already found by Legendre (1955) and Oswatitsch (1957) and shows that separation may be expected in flows with adverse pressure gradients ($p_x > 0$) and decreasing shear stress ($\tau_x < 0$).

Moreover, Eq. (3.11) shows that small separation angles imply small transverse pressure gradients $|p_y| \ll |p_x|$, suggesting that a boundary layer approximation ($p_y = 0$) might still give reasonable results near separation with small separation angles (Van Ingen 1975).

2.3 Elementary singular points in the flow (free singularities)

Returning to Eq. (3.8), free singularities in plane viscous flow patterns are easy to analyse. From the continuity equation (3.1) it follows that (3.8) is a Hamiltonian system: $\dot{x} = H_y$, $\dot{y} = -H_x$; H(x,y), the Hamiltonian, is actually equal to the familiar stream function ψ. Streamlines in the flow pattern can be seen as level curves: H = constant on the surface H = H(x,y) in the H,x,y-space. Singular points in the streamline pattern appear if u = v = 0, thus if $H_x = H_y = 0$. The character of a singular point follows in the usual manner from the eigenvalues $\lambda_{1,2}$ of the coefficient matrix of the linear part:

$$\begin{pmatrix} H_{yx} & H_{yy} \\ -H_{xx} & -H_{xy} \end{pmatrix}$$

in the singular point.

Then one finds $\lambda_1 + \lambda_2 = H_{yx} - H_{xy} = 0$ and $\lambda_1 . \lambda_2 = H_{xx}H_{yy} - H_{xy}^2$. Elementary singular points appear if $\lambda_1 . \lambda_2 \neq 0$, they will be hyperbolic *saddles* if the condition $H_{xx}H_{yy} - H_{xy}^2 < 0$ is satisfied in the singularity.

In the case $H_{xx}H_{yy} - H_{xy}^2 > 0$ the singular point is a center point, at least for the locally linearized system. But the singularity is also a *center point* for the nonlinear system since

$H_{xx}H_{yy} - H_{xy}^2 > 0$ implies that the H(x,y) surface has a local extremum at the singularity so that the level curves H = constant appear as closed curves, at least in a neighbourhood of the singularity.
So we may conclude now that the elementary free singularities either appear as a saddle or as a center.
In addition, note that in the special case of an *inviscid irrotational* flow, where the streamfunction $\psi(x,y)$ and thus H(x,y) satisfies Laplace's equation, the eigenvalue product $\lambda_1.\lambda_2$ is negative throughout and consequently only saddle points will appear in such flows.

3. Higher-order singularities in the flow pattern

Higher-order singular points in the flow pattern of a viscous fluid may be found in the interior of the flow as well on a wall. To classify the degenerate flow patterns that can occur in a viscous flow it is useful to make a clear distinction between points in the flow and points on the wall. Let us start with singularities in the flow away from the wall.

3.1 Higher-order singular points in the flow field

From the continuity equation (3.1) it follows that the flow pattern of a steady 2D viscous fluid can be described by a Hamiltonian system

$$\dot{x} = H_y, \quad \dot{y} = -H_x$$

where H(x,y) is analytic and actually equal to the streamfunction ψ.
Assume that the system has a higher-order singular point at (0,0), so that the conditions

$$H_x = H_y = 0$$

and

$$\lambda_1 . \lambda_2 = H_{xx}H_{yy} - H_{xy}^2 = 0$$

are satisfied simultaneously at this point.
To study its topological properties let the Hamiltonian system be expanded as follows

$$\dot{x} = (H_{yx})\, x + (H_{yy})\, y + P_2(x,y)$$

$$\dot{y} = -(H_{xx})\, x - (H_{xy})\, y + Q_2(x,y)$$

where the second-order partial derivatives of $H(x,y)$ are taken in (0,0) and where $P_2(x,y)$ and $Q_2(x,y)$ contain terms not lower than second degree and obey the relation

$$\frac{\partial P_2}{\partial x} + \frac{\partial Q_2}{\partial y} = 0$$

Unless H_{xx} and H_{yy} are both zero, a nonsingular linear transformation exists which transforms the system into the equivalent form

$$\dot{x} = y + P_2(x,y)$$

$$\dot{y} = Q_2(x,y)$$

and we may apply Andronov's classification scheme for higher-order singular points having both eigenvalues zero (see Chapter I). As a result we obtain the following degenerate flow patterns:

$\frac{\partial^2 Q_2}{\partial x^2} \neq 0$: cusp point singularity

$\frac{\partial^2 Q_2}{\partial x^2} = 0\,, \quad \frac{\partial^2 Q_2}{\partial x \partial y} + \frac{\partial^3 Q_2}{\partial x^3} > 0$: third-order saddle point

$\frac{\partial^2 Q_2}{\partial x^2} = 0\,, \quad \frac{\partial^2 Q_2}{\partial x \partial y} + \frac{\partial^3 Q_2}{\partial x^3} < 0$: center point

$\frac{\partial^2 Q_2}{\partial x^2} = \frac{\partial^2 Q_2}{\partial x \partial y} + \frac{\partial^3 Q_2}{\partial x^3} = 0$: flow patterns with a higher degree of degeneracy, that remain unspecified here

Qualitative sketches of the cusp point and the third-order saddle point are shown in Fig. 3.2.

Fig. 3.2. Higher order singularities away from the wall.

If H_{xx} and H_{yy} are both zero, then the Hamiltonian system has a vanishing linear part in the higher order singular point; the flow patterns have a higher degree of degeneracy and remain undiscussed.

3.2 Higher-order singular points on the wall

In paragraph 2, it was shown that the steady flow pattern near a wall is described by an autonomous dynamical system, Eq. (3.10), containing several unknown coefficients (a_1, a_2, a_3, ... etc.). Moreover, if $a_1 = 0$ and $a_2 \neq 0$ this system has a hyperbolic singularity at the origin representing the ordinary separation or attachment phenomenon.
For $a_2 = 0$ the classical Oswatitsch-Legendre solution near separation or attachment fails as a consequence of the nonhyperbolic character of the singular point. Moreover the separation angle, predicted by this solution Eq. (3.11) becomes zero.
For $a_2 = 0$, a higher-order singularity appears on the wall; a degenerate flow pattern is formed where shear stress and streamwise shear stress gradient vanish simultaneously. The streamline pattern near such a singular point satisfies the nonlinear system

$$\begin{aligned} \dot{x} &= a_3 y + a_4 x^2 + a_5 xy - \frac{2}{3} a_4 y^2 + a_7 x^3 + a_8 x^2 y - \frac{1}{2} a_7 xy^2 - \frac{1}{3} a_8 y^3 + O(4) \\ \dot{y} &= - a_4 xy - \frac{1}{3} a_5 y^2 - \frac{3}{2} a_7 x^2 y - \frac{2}{3} a_8 xy^2 + \frac{1}{2} a_7 y^3 + O(4) \end{aligned} \tag{3.12}$$

The shear stress distribution along the wall near the singular point (0,0)

$$\tau = \mu(a_4x^2 + a_7x^3 + O(x^4))$$

shows that at the singular point the distribution of the shear stress τ attains an extreme value for $a_4 \neq 0$ and an inflexion point for $a_4 = 0$, $a_7 \neq 0$.

For the case $a_4 \neq 0$, the appearance of a shear stress extremum indicates that flow reversal at the wall surface is not to be expected. The flow remains attached to the wall regardless of the occurrence of a particular point where the shear stress and streamwise shear stress gradient vanish.

In the case of $a_4 = 0$, $a_7 \neq 0$ the shear stress changes sign indicating that flow reversal may be expected. The simultaneous occurrence of an inflexion point in the shear stress distribution at the separation point suggests separations of higher-order nature.

In order to obtain a catalogue of degenerate flow patterns near a wall, let us analyse the nonlinear system (3.12) near the origin.

This system has a nonhyperbolic point at (0,0) with a double zero eigenvalue. To analyse the various possibilities of degenerate flows Eq. (3.12) will be recast into one of the two following forms.

A. $a_3 \neq 0$; nonzero streamwise pressure gradient

$$\begin{aligned} \dot{x} &= y + \hat{a}_4x^2 + \hat{a}_5xy - \frac{2}{3}\hat{a}_4y^2 + \hat{a}_7x^3 + \hat{a}_8x^2y - \frac{1}{2}\hat{a}_7xy^2 - \frac{1}{3}\hat{a}_8y^3 + O(4) \\ \dot{y} &= -\hat{a}_4xy - \frac{1}{3}\hat{a}_5y^2 - \frac{3}{2}\hat{a}_7x^2y - \frac{2}{3}\hat{a}_8xy^2 + \frac{1}{2}\hat{a}_7y^3 + O(4) \end{aligned} \tag{3.13}$$

where $\hat{a}_i = a_i\, a_3^{-1}$

B. $a_3 = 0$; zero streamwise pressure gradient

$$\begin{aligned} \dot{x} &= a_4x^2 + a_5xy - \frac{2}{3}a_4y^2 + a_7x^3 + a_8x^2y - \frac{1}{2}a_7xy^2 - \frac{1}{3}a_8y^3 + O(4) \\ \dot{y} &= -a_4xy - \frac{1}{3}a_5y^2 - \frac{3}{2}a_7x^2y - \frac{2}{3}a_8xy^2 + \frac{1}{2}a_7y^3 + O(4) \end{aligned} \tag{3.14}$$

Note the difference between the two forms: for nonzero pressure gradient ($a_3 \neq 0$) the system has a linear part whereas for zero pressure gradient ($a_3 = 0$), a system with at least quadratic terms results. Both forms will be investigated below.

Case A: Singularities with a nonzero streamwise pressure gradient ($a_3 \neq 0$)

For a nonzero pressure gradient Eq. (3.13) results which has a linear part of the form $\begin{vmatrix} 0 & 1 \\ 0 & 0 \end{vmatrix}$. The character of the nonhyperbolic point may be analysed by using Andronov's classification for degenerate points with a double zero eigenvalue (Chapter I).

On the isocline of vertical directions ($\dot{x} = 0$) $\dot{y}$ will be expanded as $\dot{y} = \psi(x) = \Delta_k x^k + o(x^k)$ where x^k is the first nonvanishing term in the expansion

$$\psi(x) = \hat{a}_4^2 x^3 + \hat{a}_4 \left(\frac{5}{2}\,\hat{a}_7 - \frac{4}{3}\,\hat{a}_4\hat{a}_5\right) x^4 + \left(\frac{3}{2}\,\hat{a}_7^2 + \hat{a}_4(\ldots)\right) x^5 + O(x^6)$$

The expansion for $\psi(x)$ points out that higher-order singularities at nonzero pressure gradient may be distinguished, corresponding to $\hat{a}_4 \neq 0$ and $\hat{a}_4 = 0$.

A1: Third-order topological saddle point ($\hat{a}_4 \neq 0$), $\tau_{xx}(0,0) \neq 0$

For $\hat{a}_4 \neq 0$ we have $k = 3$, $\Delta_k = \hat{a}_4^2$ so that the degenerate singularity is a topological saddle point of the third order.

The local flow near the third-order saddle point is described by the system

$$\dot{x} = y + \hat{a}_4 x^2$$

$$\dot{y} = -\hat{a}_4 xy$$

A sketch of the flow pattern is shown in Figure 3.3. The saddle point has four separatrices, two of them coincide with the wall and two separatrices are tangent to the wall, forming a parabolic curve. If $\tau_{xx} > 0$ this parabolic curve lies in the lower half plane $y < 0$ indicating that the flow in the upper half plane remains attached to the wall, see Figure 3.3a. On the other hand if $\tau_{xx} < 0$, the parabolically curved separatrices lie in the upper half plane so that a flow pattern results as shown in Figure 3.3b.

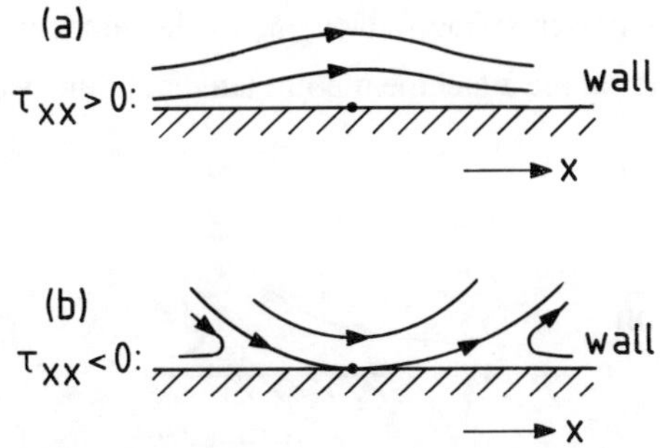

Fig. 3.3. Third order saddle points, ($p_x \neq 0$, $\tau_{xx} \neq 0$)
(a) one hyperbolic sector; (b) three hyperbolic sectors

A2: Fifth-order topological saddle point ($\hat{a}_4 = 0$), $\tau_{xx}(0,0) = 0$

For $\hat{a}_4 = 0$, we observe that the first nonvanishing term of $\psi(x)$ is of the fifth degree ($k = 5$) having a positive coefficient $\Delta_k = \frac{3}{2}\,\hat{a}_7^2$ if $\hat{a}_7 \neq 0$.

According to Andronov's classification scheme, the degenerate singularity appears to be a topological saddle point of the fifth order.

The local flow near this point is adequately described by the system

$$\dot{x} = y + \hat{a}_7 x^3$$

$$\dot{y} = \frac{3}{2}\,\hat{a}_7 x^2 y$$

The condition $\hat{a}_4 = 0$ indicates that the second derivative of the shear stress τ_{xx} vanishes at the singularity. The separatrices are found as $y = -\frac{3}{2}\,\hat{a}_7 x^3 + O(x^4)$ and $y = 0$.

Corresponding flow patterns are shown in Figure 3.4. The case $\hat{a}_7 < 0$ represents a flow separation phenomenon as sketched in Figure 3.4a. The separating streamline $y = -\frac{3}{2}\,\hat{a}_7 x^3$ lies in the upper half plane, leaving the wall tangentially in streamwise direction.

The other case $\hat{a}_7 > 0$ represents flow attachement; the separatrix is an attachment streamline which terminates at the attachment point tangent to the wall (see Fig. 3.4b).

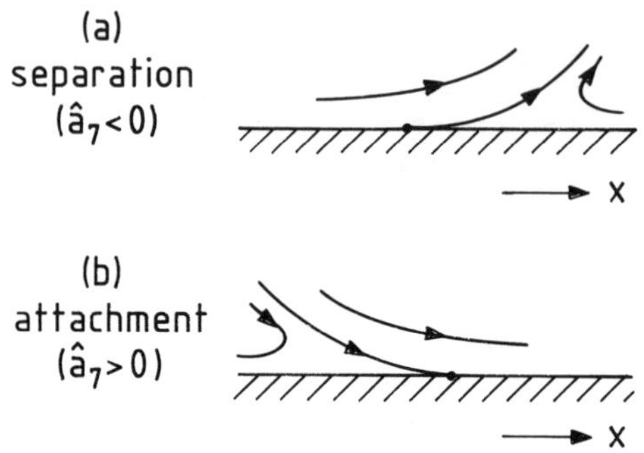

Fig. 3.4. Fifth-order saddle points. ($p_x \neq 0$, $\tau_{xx} = 0$)

The study of the case $\hat{a}_7 = 0$, is not taken up in this work, a further expansion of the velocity components would then be necessary.

Case B: Singularities with zero pressure gradient ($a_3 = 0$)

Singular flow patterns, having zero pressure gradient ($a_3 = 0$) in streamwise direction are described by the system (3.14):

$$\begin{aligned} \dot{x} &= a_4x^2 + a_5xy - \frac{2}{3}a_4y^2 + a_7x^3 + a_8x^2y - \frac{1}{2}a_7xy^2 - \frac{1}{3}a_8y^3 + O(4) \\ \dot{y} &= -a_4xy - \frac{1}{3}a_5y^2 - \frac{3}{2}a_7x^2y - \frac{2}{3}a_8xy^2 + \frac{1}{2}a_7y^3 + O(4) \end{aligned} \tag{3.14}$$

Since the linear part is missing, the classification of Andronov cannot be used in obtaining the degenerate flow patterns. So we are left with the basic question: which terms in the right-hand side of (3.14) are essential for the local topology of the degenerate state?

We will prove that the quadratic part of (3.14) determines the qualitative pattern provided that $a_4 \neq 0$.

The main tool is a technique called 'blowing-up', for example used by Andronov (1973) and Takens (1974) to study degenerate singularities with at least one linear term. It will be shown here that the blowing-up method also suffices if the linear part is missing. The idea consists of introducing polar coordinate transformations to expand degenerate singular points into curves containing a finite number of singular points. If these are hyperbolic after blow-up then the local phase portrait near the curve, and hence near the *original* degenerate point, is unamenable to the influence of higher-order terms.

In the case of Eq. (3.14) one blow-up, provided $a_4 \neq 0$, is sufficient to obtain a stable transformed vector field.

Here we describe this blow-up in detail to illustrate the method.

Letting $x = r\cos\theta$, $y = r\sin\theta$ $(r \geq 0, 0 \leq \theta \leq \pi)$ we get Eq. (3.14) recast in the polar coordinate form

$$\begin{aligned} \dot{r} &= r(a_4c^3 + a_5c^2s - \tfrac{5}{3}a_4cs^2 - \tfrac{1}{3}a_5s^3) + O(r^2) \\ \dot{\theta} &= -2a_4sc^2 - \tfrac{4}{3}a_5cs^2 + \tfrac{2}{3}a_4s^3 + s\,.\,O(r) \end{aligned} \tag{3.15}$$

where c and s are shorthand notations for $\cos\theta$ and $\sin\theta$ respectively.

We now expand the singularity at $r = 0$ by regarding (r,θ) as a cartesian coordinate system. On the interval $0 \leq \theta \leq \pi$ (3.15) has singular points on $r = 0$ at $\theta_1 = 0$, $\theta_2 = \pi$ and $\theta_{3,4} = \tan^{-1}(\hat{a}_5 \pm \sqrt{3 + \hat{a}_5^2})$, where $\hat{a}_5 = a_5/a_4$.

An eigenvalue calculation shows that all these points are hyperbolic provided that $a_4 \neq 0$. Moreover, it appears that the eigenvalue product is negative througout, implying that these points are saddle points.

For $a_4 \neq 0$ the 'blow-up' pattern is shown in Figure 3.5 (arrows denote flow directions which correspond to the case $a_4 > 0$).

Since for $a_4 \neq 0$ all singular points are hyperbolic after the first blow-up, the flow of Eq. (3.15) is stable with respect to small (higher-order) perturbations. Consequently, a similar

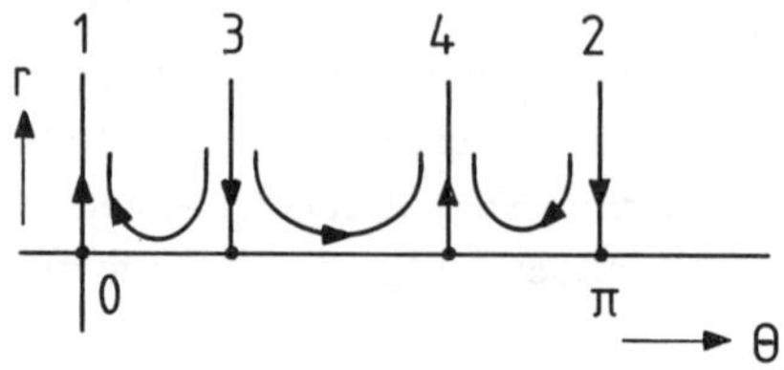

Fig. 3.5. First blow-up of Eq. 3.14 (a_4>0).

conclusion follows for Eq. (3.14); the degenerate flow problem is determined topologically by the quadratic terms and its topological structure cannot be disturbed by adding higher-order terms.

B1: Saddle point with three hyperbolic sectors ($a_4 \neq 0$), $\tau_{xx}(0,0) \neq 0$

The flow patterns that will occur in the physical plane for $a_4 \neq 0$ may be obtained if Figure 3.5 is 'blown down' from $(r,\theta) \rightarrow (x,y)$. These patterns are drawn in Figure 3.6, the flow direction corresponds to $a_4 > 0$.

The blowing-up method evidences that the topological properties of these patterns are sufficiently described by the reduced system

$$\dot{x} = a_4x^2 + a_5xy - \frac{2}{3}a_4y^2$$

$$\dot{y} = -a_4xy - \frac{1}{3}a_5y^2$$

The degenerate flow pattern in question has a saddle point behaviour; there are four separatrices, dividing the upper half plane into three hyperbolic sectors. The separation angles α_1 and α_2 (Fig. 3.6) satisfy the relation $\tan\alpha_1 . \tan\alpha_2 = 3$.

The shear stress is positive everywhere on the wall surface except in the singular point where it reaches zero as a minimum value. The actual relationship is given by $\tau = \mu(a_4 x^2 + O(3))$.

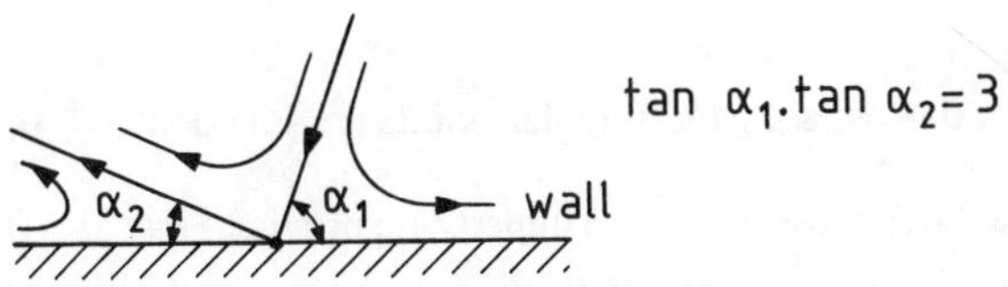

Fig. 3.6. Saddle point with three hyperbolic sectors ($p_x = 0, \tau_{xx} \neq 0$).

B2: Saddle point with two or four hyperbolic sectors ($a_4 = 0$), $\tau_{xx}(0,0) = 0$

For $a_4 = 0$ the *first* blow-up given by Eq. (3.15) reduces to

$$\begin{aligned} \dot{r} &= r(a_5 c^2 s - \frac{1}{3} a_5 s^3) \\ \dot{\theta} &= -\frac{4}{3} a_5 c s^2 + s \,.\, O(r) \end{aligned} \tag{3.16}$$

On the interval $0 \leq \theta \leq \pi$, (3.16) has three singular points on $r = 0$, they occur at $\theta = 0$, $\theta = \frac{\pi}{2}$ and $\theta = \pi$.

The singular point at $\theta = \frac{\pi}{2}$ is a hyperbolic saddle point if $a_5 \neq 0$. The points at $\theta = 0$ and $\theta = \pi$ appear to be nonhyperbolic and a second blow-up $(r,\theta) \rightarrow (\eta,\psi)$ has to be performed. Setting $\theta \approx 0$, $r \approx 0$ we expand (3.14) in Taylor series near $(r,\theta) = (0,0)$ to obtain:

$$\begin{aligned} \dot{r} &= a_5 r\theta + a_7 r^2 + O(3) \\ \dot{\theta} &= -\frac{4}{3} a_5 \theta^2 - \frac{5}{2} a_7 r\theta + O(3) \end{aligned}$$

A similar form is obtained near $(r,\theta) = (0,\pi)$.

After the second blow-up, defined by $\theta = \eta \cos\psi$, $r = \eta \sin\psi$ $(\eta > 0, 0 \leq \psi \leq \frac{\pi}{2})$ the following vector field is obtained

$$\dot{\eta} = \eta\{a_5\sin^2\psi\cos\psi + a_7\sin^3\psi - \frac{4}{3}a_5\cos^3\psi - \frac{5}{2}a_7\sin\psi\cos^2\psi\} + O(\eta^2)$$

$$\dot{\psi} = \frac{7}{3}a_5\sin\psi\cos^2\psi + \frac{7}{2}a_7\cos\psi\sin^2\psi + \sin\psi\,.\,O(\eta) \tag{3.17}$$

Within the domain $0 \le \psi \le \frac{\pi}{2}$, the singular points are found at $\eta = 0$, $\psi = 0$, $\frac{\pi}{2}$ and $\psi = \tan^{-1}(-2/3\ a_5/a_7)$. All these points are hyperbolic provided that $a_5 \neq 0$, and $a_7 \neq 0$. Hence the vector field of (3.17) is stable with respect to higher-order perturbations.

We now 'blow-down' once to get the phase portraits in the r,θ plane (Fig. 3.7). The flow direction in Figure 3.7 corresponds to $a_5 > 0$. The next 'blow-down' to the physical x,y-plane results in degenerate flow patterns as depicted in Figure 3.8. The blowing-up method shows that the flow topology of these degenerate states is fully determined by the system ($a_5 \neq 0$, $a_7 \neq 0$).

$$\dot{x} = a_5xy + a_7x^3$$

$$\dot{y} = -\frac{1}{3}a_5y^2 - \frac{3}{2}a_7x^2y$$

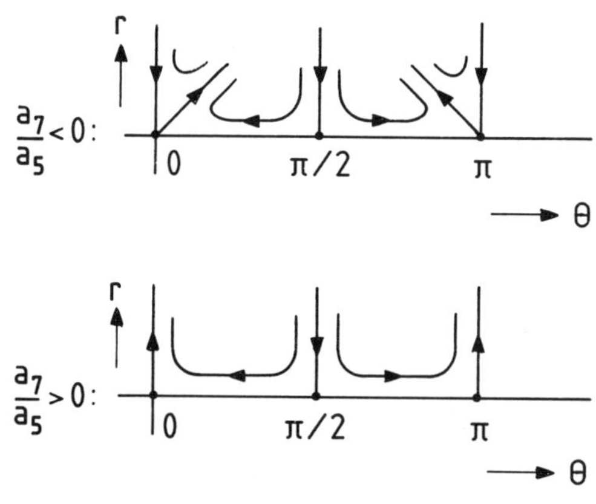

Fig. 3.7. First blow up of Eq. 3.14, ($a_4 = 0$, $a_5 \neq 0$)

The degenerate flows have a saddle nature, if $a_5a_7 > 0$ (< 0) there appear two (four) hyperbolic sectors in the upper half plane, respectively.

As we know, the condition $a_3 = a_4 = 0$ corresponds to points at the wall where the streamwise pressure gradient (p_x) and the second order shear stress gradient (τ_{xx}) vanish simultaneously. This involves that the observed flow may be discerned according to the sign of τ_{xxx} and the sort of pressure extreme (maximum or minimum). At a local maximum of the wall pressure ($a_5 < 0$) the flow separates

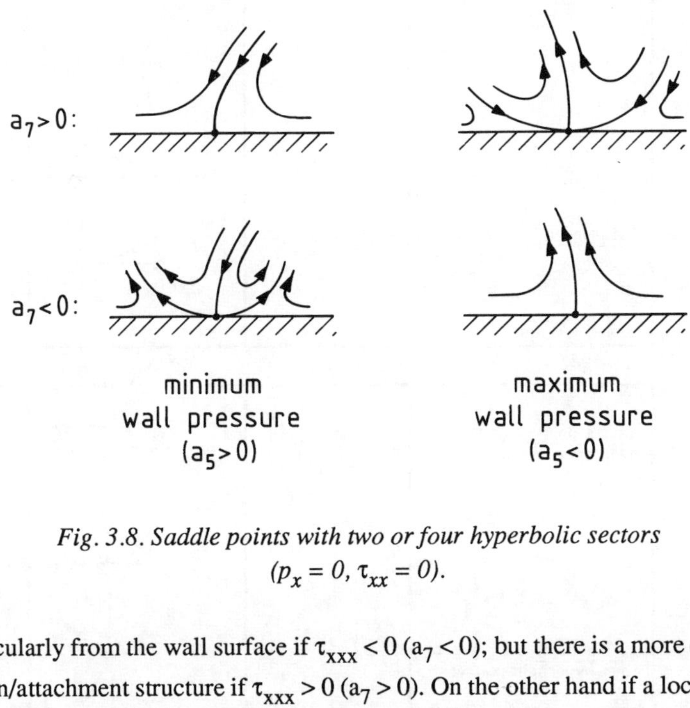

Fig. 3.8. Saddle points with two or four hyperbolic sectors ($p_x = 0$, $\tau_{xx} = 0$).

perpendicularly from the wall surface if $\tau_{xxx} < 0$ ($a_7 < 0$); but there is a more complicated separation/attachment structure if $\tau_{xxx} > 0$ ($a_7 > 0$). On the other hand if a local minimum of the wall pressure is present, the flow attaches perpendicularly to the wall if $\tau_{xxx} > 0$ and a complicated attachment/ separation pattern occurs if $\tau_{xxx} < 0$.

In conclusion the elementary and higher-order singularities on the wall, encountered so far are summarized in the next table.

	singularity	vector field	flow patterns		τ_x	τ_{xx}	τ_{xxx}	p_{xx}
Hyperbolic $\tau_x(0,0) \neq 0$	saddle point Oswatitsch-Legendre solution for elementary separation and attachment	$\begin{pmatrix} y+\hat{a}_2 x \\ -\frac{1}{2}\hat{a}_2 y \end{pmatrix}$	$\hat{a}_2<0$		−			
			$\hat{a}_2>0$	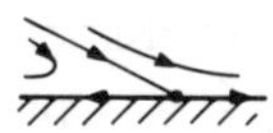	+			
Non-hyperbolic nonzero pressure gradient $\tau_x(0,0)=0$ $p_x(0,0)\neq 0$	third-order saddle	$\begin{pmatrix} y+\hat{a}_4 x^2 \\ -\hat{a}_4 xy \end{pmatrix}$	$\hat{a}_4>0$	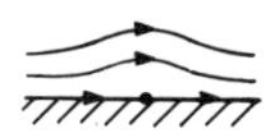	0	+		
			$\hat{a}_4<0$	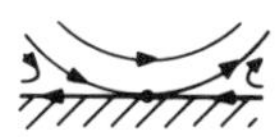	0	−		
	fifth-order saddle	$\begin{pmatrix} y+\hat{a}_7 x^3 \\ -\frac{3}{2}\hat{a}_7 x^2 y \end{pmatrix}$	$\hat{a}_7<0$		0	0	−	
			$\hat{a}_7>0$	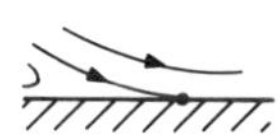	0	0	+	
Non-hyperbolic zero pressure gradient $\tau_x(0,0)=0$ $p_x(0,0)=0$	saddle point with three hyperbolic sectors	$\begin{pmatrix} x^2+\hat{a}_5 xy-\frac{2}{3}y^2 \\ -xy-\frac{1}{3}\hat{a}_5 y^2 \end{pmatrix}$			0	+		
	saddle point with two or four hyperbolic sectors	$\begin{pmatrix} a_5 xy+a_7 x^3 \\ -\frac{1}{3}a_5 y^2-\frac{3}{2}a_7 x^2 y \end{pmatrix}$	$a_5<0$, $a_7<0$	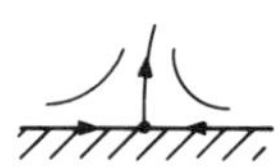	0	0	−	−
			$a_5<0$, $a_7>0$	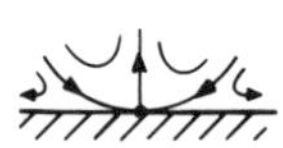	0	0	+	−
			$a_5>0$, $a_7>0$	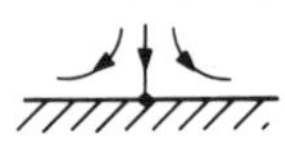	0	0	+	+
			$a_5>0$, $a_7<0$	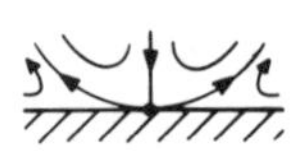	0	0	−	+

Table: Hyperbolic and nonhyperbolic singularities on the wall in two-dimensional viscous flow.

4. Unfolding of the topological saddle point of the third order

4.1 Local phase portraits of the unfolding

The steady viscous flow along a wall has a third-order topological saddle point if $\tau = \tau_x = 0$, $\tau_{xx} \neq 0$ and $p_x \neq 0$ at this point.

The streamlines form a topological pattern similar to that of the trajectories of the dynamical system (see Eq. (3.13))

$$\begin{aligned} \dot{x} &= y + \hat{a}_4 x^2 + \hat{a}_5 xy - \frac{2}{3}\hat{a}_4 y^2 + O(3) \\ \dot{y} &= -\hat{a}_4 xy - \frac{1}{3}\hat{a}_5 y^2 + O(3) \end{aligned} \tag{3.18}$$

The degenerate singularity in (0,0) has a double zero eigenvalue implying that the center eigenspace E^c coincides with the physical plane (E^s and E^u are both empty).

As a result, the application of center manifold theory provides no reduction of the dimension of the state space which contains all qualitative features of the bifurcation.

Consequently we are left with a second-order system whose unfoldings have to be considered in the x,y space. The physical unfoldings, obeying the no-slip boundary condition, follow from Eq. (3.10) by regarding the coefficients a_1 and a_2 as the bifurcation parameters both having zero as the bifurcation value. Small perturbations of a_1 and a_2 with respect to their bifurcation value will be denoted by μ_1, μ_2 respectively, so that we may write

$$\hat{a}_1 = 0 + \mu_1, \quad \hat{a}_2 = 0 + \mu_2$$

The physical unfolding of Eq. (3.18) becomes

$$\begin{aligned} \dot{x} &= \mu_1 + \mu_2 x + y + \hat{a}_4 x^2 + \hat{a}_5 xy - \frac{2}{3}\hat{a}_4 y^2 + O(3) = P(x,y\,;\mu_1,\mu_2) \\ \dot{y} &= -\frac{1}{2}\mu_2 y \qquad -\hat{a}_4 xy - \frac{1}{3}\hat{a}_5 y^2 + O(3) = Q(x,y;\mu_1,\mu_2) \end{aligned}$$

The phase plane portraits of the unfolding are dominated by the singular points (x_i,y_i) which might appear after bifurcation. The locations of these points are obtained as those solutions of

$$P(x_i, y_i; \mu_1, \mu_2) = 0$$
$$Q(x_i, y_i; \mu_1, \mu_2) = 0 \tag{3.19}$$

which satisfy the condition

$$\lim_{\substack{\mu_1 \to 0 \\ \mu_2 \to 0}} (x_i, y_i) = (0,0)$$

The solutions of (3.19) $x_i = x_i(\mu_1,\mu_2)$ and $y_i = y_i(\mu_1,\mu_2)$ may be represented by surfaces. Since μ_1 and μ_2 are assumed to be small we are only interested in the local geometry of the surfaces near the origin. This enables us to use a parametrisation technique by means of which the qualitative properties of these surfaces can be fully understood.

Using this technique one looks in fact for the leading terms in Eq. (3.19) which determine the position of singular points as functions of the bifurcation parameters.
Let $x_i = \bar{x}_i \mu_2$, $y_i = \bar{y}_i \mu_2^2$, $\mu_1 = k \mu_2^2$, then (3.19) becomes

$$(k + \bar{x}_i + \bar{y}_i + \hat{a}_4 \bar{x}_1^2)\, \mu_2^2 + O(\mu_2^3) = 0$$

$$(-\frac{1}{2}\bar{y}_i - \hat{a}_4 \bar{x}_i \bar{y}_i)\, \mu_2^3 + O(\mu_2^4) = 0$$

which shows that the leading terms for x_i and y_i may be obtained from

$$\mu_1 + \mu_2 x_i + y_i + \hat{a}_4 x_i^2 = 0$$

$$-\frac{1}{2}\mu_2 y_i - \hat{a}_4 x_i y_i = 0$$

These equations may be solved for x_i and y_i to obtain the following singular points

$$S_{1,2}:\left(\frac{-\mu_2 \pm \sqrt{\mu_c}}{2\hat{a}_4}, 0\right) \qquad \text{on the wall}$$

$$C:\left(-\frac{\mu_2}{2\hat{a}_4}, \frac{\mu_c}{4\hat{a}_4}\right) \qquad \text{away from the wall}$$

where μ_c is a compound bifurcation parameter: $\mu_c = \mu_2^2 - 4\hat{a}_4\mu_1$.

Let us consider the two cases $\hat{a}_4 > 0$ and $\hat{a}_4 < 0$ in the semi-plane $y \geq 0$ separately, taking $\hat{a}_4 > 0$ first.

Case $\hat{a}_4 > 0$: third-order saddle point with one hyperbolic sector

For $\mu_c<0$ no singular points occur in the semi-plane $y \geq 0$; the original degenerate singularity disappears and the flow remains attached to the wall.

For $\mu_c > 0$ there appear three distinct singular points in the upper half plane. The points S_1 and S_2 are attached to the wall and both are saddle points; above the wall and lying midway between these saddle points, a center point C is found.

Away from the wall we also observe a saddle-to-saddle connection between S_1 and S_2 enclosing the center point C. A short calculation shows that this connection is given by the equation

$$y = -\frac{3}{2}\hat{a}_4 \{(x + \frac{1}{2}\frac{\mu_2}{\hat{a}_4})^2 - \frac{\mu_c}{4\hat{a}_4^2}\}$$

In the particular case $\mu_c = 0$, the three singular points collapse into a single point on the wall ($x = -\mu_2/2\hat{a}_4$), in which the degenerate singularity, being the original third-order topological saddle point, remains. It means that parameter changes satisfying $\mu_c = 0$ do not remove nor unfold the original degeneracy; consequently $\mu_c = 0$ represents a nongeneric perturbation.

The bifurcation set, together with the possible flow patterns of the unfolding of Eq. (3.18) are shown in Figure 3.9a.

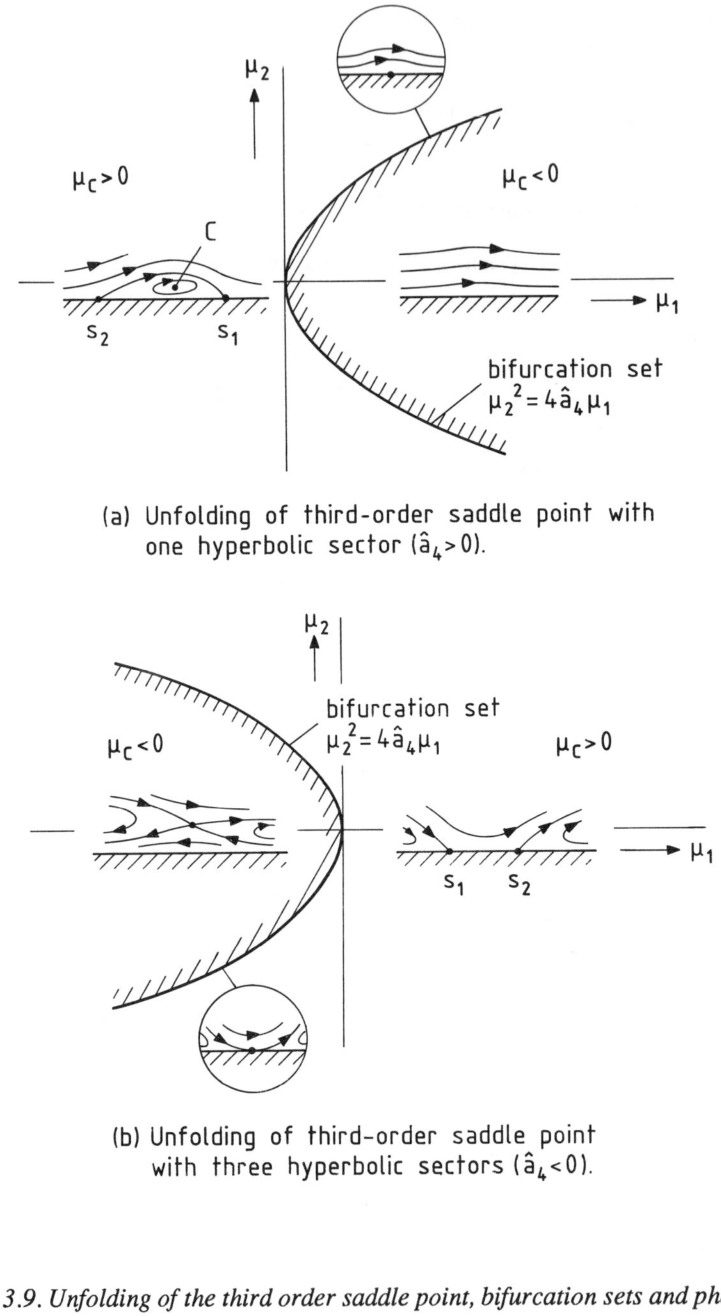

Fig. 3.9. Unfolding of the third order saddle point, bifurcation sets and phase portraits.

Case $\hat{a}_4 < 0$: third-order saddle point with three hyperbolic sectors

The unfolding of this degenerate singularity is shown in Figure 3.9b. For $\mu_c > 0$ there appear only singular points on the wall. Two distinct saddle points S_1 and S_2 are observed there, resulting in a flow pattern where flow attachement (S_1) is followed by flow separation (S_2).

For $\mu_c < 0$ there is only one saddle point in the flow away from the wall and the local flow structure in the interface between opposing flow directions results.

The case $\mu_c = 0$ represents a nongeneric bifurcation, so that the original higher-order singularity remains as a structurally unstable element in the flow.

4.2 Incipient bubble separation

Let us consider the unfolding of the third-order saddle point having one hyperbolic sector ($\hat{a}_4 > 0$) in more detail.

A generic perturbation of the third-order saddle point can give, at least locally, either a flow fully attached to the wall or a flow pattern with separation and reattachment. The latter is recognized as the well known two-dimensional laminar separation bubble.

Separation bubbles are features in laminar flows that have been studied for many years and it is well-known that near both the separation point and the reattachment point the flow satisfies the Oswatisch-Legendre condition as given by Eq. (3.11). However, in this study the separation bubble also appears in a local solution of the Navier-Stokes equations showing a coherent flow structure including the essential features as separation point, reattachment point, recirculating region, vortex center and dividing streamline. Moreover, this flow structure arises as a bifurcation solution of a nonhyperbolic singularity so that the genesis of separation bubbles can be described in terms of an unfolding of this singularity.

A necessary condition for incipient bubble-type separation can now be derived. It is $\mu_c > 0$, or expressed in terms of wall shear stress:

$$\tau_x^2 - 2\tau\, \tau_{xx} > 0$$

This condition points out that separation bubbles arise in the flow near a point where $\tau = \tau_x = 0$ if the flow is disturbed so that $\tau_x^2 - 2\tau\tau_{xx} > 0$ in this point.

The bifurcation solution for bubble-type separation yields information on the geometry of small separation bubbles. Actually we can calculate the shape of the bubble for the limiting condition $\mu_c \to 0$. The leading terms for the height of the bubble h_b and for the height of the vortex center h_c above the wall with respect to the bubble length l become:

bubble height $\quad h_b/l = \frac{3}{8}\sqrt{\mu_c}$

vortex center $\quad h_c/l = \frac{2}{8}\sqrt{\mu_c}$

Expressed in terms of separation angle θ_s we obtain

bubble height $\quad h_b/l = \frac{1}{4}\tan\theta_s$

vortex center $\quad h_c/l = \frac{1}{6}\tan\theta_s$

The first expression shows that bubbles in embryonic state are very slender, in the sense that they are more extended in flow direction than in the direction normal to the wall. Furthermore, the center point of the vortex always lies in the upper part of the recirculating region.

4.3 Separation along a moving wall

In this section we address the problem to describe flow separation near a wall moving in its own plane as an unfolding of the third-order topological saddle point having one hyperbolic sector, $\hat{a}_4 > 0$. The set-up of this problem is basically the same as given for the fixed-wall case with the exception that the wall moves with a constant wall velocity u_w in its own plane. Positive values of u_w mean that the wall moves in downstream direction whereas negative values of u_w represent an upstream moving wall.

Separated flow patterns near the moving wall will be investigated by appropriate unfolding the third-order topological saddle point (Eq. 3.13) so that the unfolding satisfies the boundary conditions

$$u = u_w,\ v = 0 \quad \text{on the wall.}$$

In order to obtain these unfoldings system (3.10), which is used in the fixed wall case, has to be replaced by

$$\begin{aligned} \dot{x} &= u_w + y(a_1+a_2x+a_3y+a_4x^2+a_5xy+a_6y^2) + O(4) \\ \dot{y} &= y(-\frac{1}{2}a_2y-a_4xy-\frac{1}{3}a_5y^2) + O(4) \end{aligned} \tag{3.20}$$

which satisfies the continuity equation (3.1).
The vorticity transport equation, Eq. (3.3) requires:

$$2a_4 + 3a_6 = \frac{u_w a_2}{2\nu}$$

Note that (3.20) reduces to (3.10) if $u_w = 0$.

Let us examine the phase portraits (flow patterns) of Eq. (3.20) near the origin (0,0), which is a nonhyperbolic third-order saddle point if $a_1 = a_2 = u_w = 0$. To obtain flow patterns valid near a moving wall we take, in addition to a_1 and a_2, also u_w as a bifurcation parameter having a bifurcation value $u_w = 0$. Small variations with respect to this bifurcation value will be denoted by $\mu_o = u_w/a_3$.
Consulting Eq. (3.20) we find that we have to study the three-parameter unfolding

$$\begin{aligned} \dot{x} &= \mu_o + y[\mu_1+\mu_2x+y+\hat{a}_4x^2+\hat{a}_5xy - [\frac{2}{3}\hat{a}_4 - a_3\frac{\mu_o\mu_2}{6\nu}]\,y^2 + O(3)] = P(x,y) \\ \dot{y} &= y^2[-\frac{1}{2}\mu_2 - \hat{a}_4x - \frac{1}{3}\hat{a}_5\,y + O(3)] = Q(x,y) \end{aligned} \tag{3.21}$$

with $\hat{a}_i = a_i(a_3)^{-1}$ and where μ_o, μ_1 and μ_2 are bifurcation parameters.
Singular points (x_i,y_i) of (3.21) which appear near (0,0) after bifurcation will be found by using the parametrisation technique adopted in paragraph (4.1) for the fixed wall case. Referring to this case we propose

$$x_i = \bar{x}_i\mu_2,\ y_i = \bar{y}_i\mu_2^2,\ \mu_o = k_o\mu_2^4,\ \mu_1 = k_1\mu_2^2$$

then

$$P(x_i, y_i) = 0,$$

$$Q(x_i, y_i) = 0$$

becomes

$$(k_o + k_1\bar{y}_i + \bar{x}_i\bar{y}_i + \bar{y}_i^2 + \hat{a}_4\bar{x}_i^2\bar{y}_1)\ \mu_2^4 + O(\mu_2^5) = 0$$

$$(-\bar{y}_i^2 - \hat{a}_4\bar{x}_i\bar{y}_i^2)\ \mu_2^5 + O(\mu_2^6) = 0$$

which shows that the leading terms for x_i and y_i follow from

$$\mu_o + \mu_1 y_i + \mu_2 x_i y_i + y_i^2 + \hat{a}_4 x_i^2 y_i = 0$$

$$(\frac{1}{2}\mu_2 + \hat{a}_4 x_i)\ y_i^2 = 0$$

This system may be solved for x_i and y_i to obtain two saddle points:

$$S_{1,2}:\ \frac{-\mu_2 \pm \sqrt{\mu_c}}{2\hat{a}_4}, 0 \quad ;\ \text{on the wall if } \mu_o = 0 \text{ (fixed wall)}$$

and one saddle point

$$S_3:\ -\frac{\mu_2}{2\hat{a}_4},\ \frac{\mu_c - \sqrt{\mu_c^2 - \mu_o(8\hat{a}_4)^2}}{8\hat{a}_4},\ \text{away from the wall if } \mu_o \neq 0 \text{ (moving wall)}$$

Furthermore a center point can appear:

$$C:\ -\frac{\mu_2}{2\hat{a}_4},\ \frac{\mu_c + \sqrt{\mu_c^2 - \mu_o(8\hat{a}_4)^2}}{8\hat{a}_4}$$

with $\mu_c = \mu_2^2 - 4\hat{a}_4\mu_1$; the compound bifurcation parameter already encountered in the fixed-wall solution.

Note that the bifurcation looks a bit unfamiliar. The free saddle point S_3 moves to the position $(-\mu_2/2\hat{a}_4, 0)$ on the wall if $\mu_o \to 0^+$, but this position is not a singular point if $\mu_o = 0$. On the other hand two adjacent saddle points S_1 and S_2 are present on the wall only if $\mu_o = 0$; the singular character of these points is removed if μ_o is taken nonzero.
The unusual bifurcation is explained if system (3.20) is understood to unfold a line of singularities, lying at y = 0 and appearing in the vector field $\dot{x} = y^2 + \hat{a}_4 x^2 y$, $\dot{y} = -\hat{a}_4 y^2 x$.

The location of the singular points is essentially determined by the compound bifurcation parameter μ_c and the wall velocity μ_o which then appear to describe the unfolding completely.

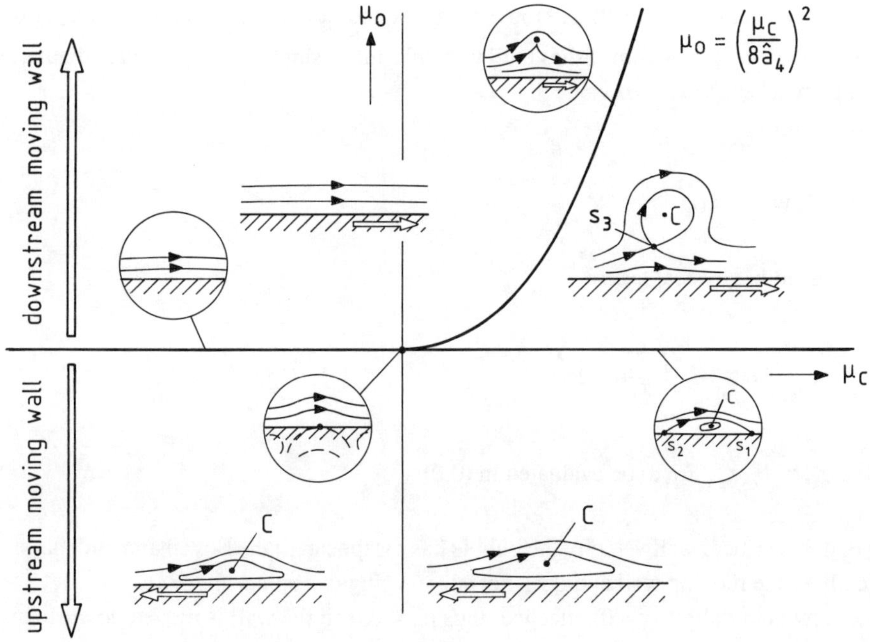

Fig. 3.10. Flow patterns near moving walls.

For fixed μ_c and μ_o a variation of μ_2 results only in a shift $\mu_2/2\hat{a}_4$ in x-direction of the singular points and thus a shift of the complete flow structure.
Bifurcation sets and phase portraits in the upper half plane are displayed in Figure 3.10.

For $\mu_o = 0$ the fixed-wall case emerges as a particular solution, representing a bubble-type separation if $\mu_c > 0$ and attached flow conditions if $\mu_c < 0$.
Moving the wall will affect these flows as follows (see Figure 3.10).
Let us start with the case that the flow is initially ($\mu_o = 0$) separated ($\mu_c > 0$). Moving the wall downstream ($\mu_o > 0$) will at once remove the saddle points S_1 and S_2 and will generate a free saddle point S_3 and a center point C in the flow. The closed loop formed by the separatrices above S_3 is a streamline encircling the closed streamlines around the center point C, forming a region with recirculating flow above the wall. The lower separatrices of S_3 extend far upstream and downstream, dividing the flow in an upper and a lower part. The lower part accomodates with the moving wall and flows underneath the recirculating bubble; the upper part of the flow passes over the bubble. A further increase of the wall velocity at constant μ_c ($\mu_c > 0$) will move the center point C and the saddle point S_3 closer together, the bubble size diminishes until it shrinks into a single nonhyperbolic singularity when μ_o reaches the critical value

$$\mu_{o_{crit.}} = \left(\frac{\mu_c}{8\hat{a}_4}\right)^2 \tag{3.22}$$

or expressed in physical terms

$$u_{w_{crit.}} = \frac{1}{8\mu\, p_x}\left(\frac{\tau_x^2}{\tau_{xx}} - 2\tau\right)^2$$

where p_x, τ, τ_x, τ_{xx} are to be evaluated in (0,0).

Beyond this critical wall velocity the bubble has disappeared; the flow pattern adapts sufficiently to the moving wall and does not contain singular points anymore.
If the flow is initially ($\mu_o = 0$) attached, thus $\mu_c < 0$, and the wall is moved downstream, $\mu_o > 0$, the unfolding shows that the attached flow conditions are maintained.
Consequently, unfolding theory reflects the well-known principle that separation near a wall will be prevented if the wall is moving downstream at a sufficiently high speed.

Next, if the wall is moved upstream ($\mu_o < 0$), irrespective of the value of μ_c a flow pattern is observed having a single center point above the wall. This center point is part of a recirculating bubble extending from far upstream to far downstream at least according to the local approximation. To elucidate this large streamwise extension of the bubble let us observe the isoclines $\dot{x} = 0$ and $\dot{y} = 0$ in the phase portraits of Eq. (3.21). These curves are sketched in Figure 3.11 for the cases $\mu_o = 0$ (fixed wall), $\mu_o < 0$ (upstream moving wall) and $\mu_o > 0$ (downstream moving wall).

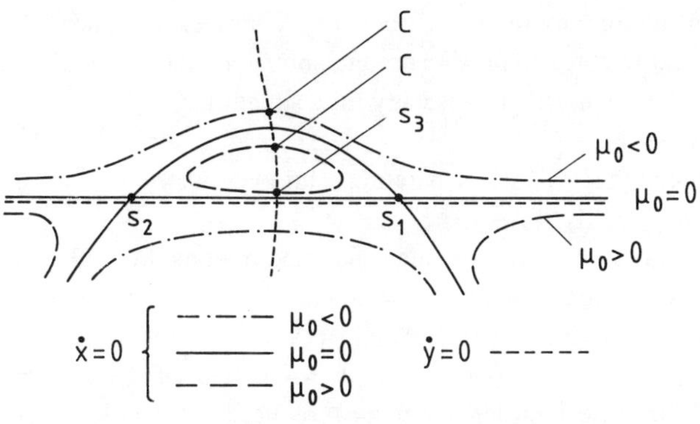

Fig. 3.11. Influence of wall velocity on isoclines $\dot{x} = 0$ and $\dot{y} = 0$.

The $\dot{x} = 0$ curve has two branches; in the fixed-wall case one branch coincides with the wall surface, the other forms a parabola intersecting the wall at S_1 and S_2. Imposing a small wall velocity ($\mu_o < 0$) will cause these branches to be perturbed as follows. Apart from a branch in the lower half plane there appears an $\dot{x} = 0$ curve in the upper half plane, lying above the unperturbed position and extending from far upstream to far downstream (Figure 3.11).

To the order considered, the $\dot{y} = 0$ curves ($y = 0$ and $x = -\mu_2/2\hat{a}_4$) remain unaffected by the moving wall.

The behaviour of the $\dot{x} = 0$ and $\dot{y} = 0$ curves in the upper half plane indicate the formation of a recirculating bubble close to the wall surface if the wall is moved upstream. The bubble is small in height but extends from far upstream to far downstream.

The previous observations concerning the unfolding of Eq. (3.21) seem to support the idea that:

- Moving a wall downstream can prevent or at least delay separation and bubble formation.
- Moving a wall upstream will always lead to a 'separation' of the main flow; a small recirculating layer appears underneath the main flow.

4.4 MRS-criterion for separations in flows along a moving wall

The flow patterns along moving walls, obtained by unfolding a third-order saddle point with one hyperbolic sector, will now be used to comment on the so-called MRS-criterion, developed to predict separation in unsteady flow situations.

It is generally agreed that Prandtl's separation criterion: $\partial u/\partial y = 0$ at $y = 0$, is adequate to predict separation in steady flows.
However, in unsteady flows with moving separation regions, Prandtl's criterion fails and a suggestion was made to replace it by the so called MRS-criterion developed independently by Moore (1958), Rott (1956) and Sears (1956). The MRS-criterion predicts separation in unsteady flows if the condition $\partial u/\partial y = u = 0$ is satisfied in some point of the flow field. The MRS-criterion ($\partial u/\partial y = 0$, $u=0$) reduces to Prandtl's criterion ($\partial u/\partial y = 0$, $y = 0$) if the wall ($y = 0$) is at rest in the reference frame in which u is measured.
Points in the flow field where the condition $u = 0$, $\partial u/\partial y = 0$ is fulfilled are called MRS-points.
It is not clear, however, what is to be understood precisely by separation in this context. This will be evident in the steady flow along a moving wall, this flow is unsteady in a reference system fixed to the wall, but no streamline separates from the wall.

Using the MRS-criterion ($\partial u/\partial y = 0$, $u = 0$), Sears and Telionis (1975) proposed a model for separation in the steady flow near a moving wall. This so-called MRS-model is presented in Figure 3.12, where streamline patterns and velocity distributions u(y) are shown. Irrespective of whether the wall moves upstream or downstream, the MRS-model reveals the formation of a single saddle point in the flow away from the moving wall. The velocity distributions u(y) indicate that this saddle point is a MRS-point, satisfying the MRS-criterion $\partial u/\partial y = 0$, $u = 0$.

Several numerical and experimental attempts have been made to verify the MRS-model together with the corresponding velocity profiles by considering the steady flow along a moving wall. A concise and profound review of most of these attempts is given by Williams (1977). We do not feel the need to summarize them here in detail, apart from some remarks supporting William's conclusion that the MRS-model is applicable for the downstream moving case; for upstream moving walls the question is still open.

Concerning the applicability in the *downstream case*, the contributions of Telionis and Werlé (1973), Tsahalis and Telionis (1973) and Danberg and Fansler (1975) have to be mentioned in particular. Their results are based on the numerical solution of the boundary layer equations and show a close agreement of the calculated velocity profiles with those expected from the MRS-model.

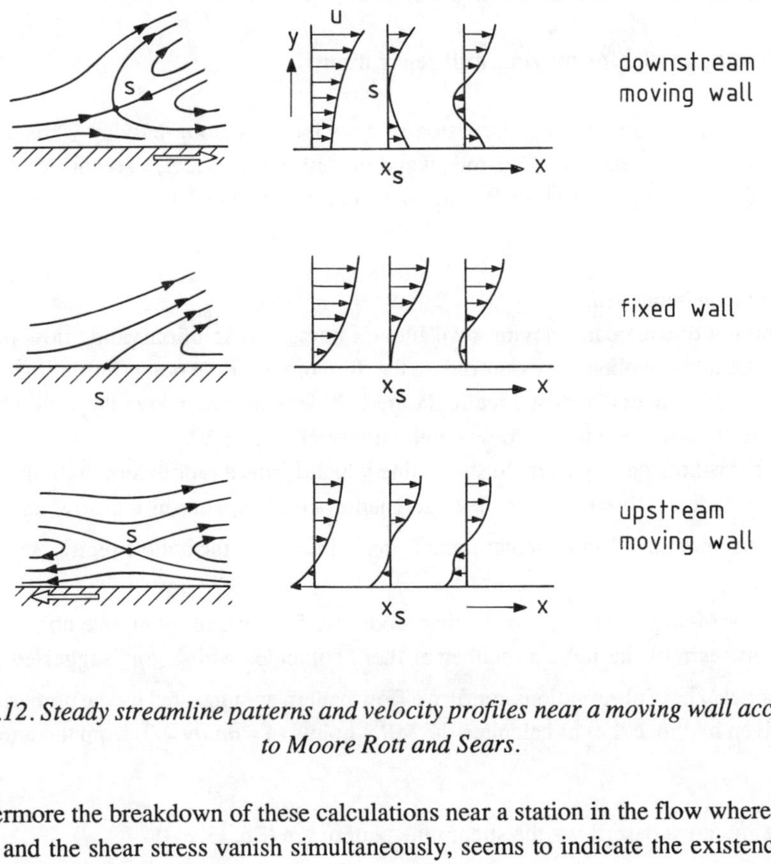

Fig. 3.12. Steady streamline patterns and velocity profiles near a moving wall according to Moore Rott and Sears.

Furthermore the breakdown of these calculations near a station in the flow where the velocity and the shear stress vanish simultaneously, seems to indicate the existence of an MRS-point in the flow.

More recently, Inoue (1981) presented a numerical study of the steady flow along a moving wall. Instead of using the classical boundary layer equations, he solves the Navier Stokes equations approximately by neglecting only the diffusion terms in the main flow direction. This approximate version of the NS-equations has some advantages in comparison with the boundary layer equations. First, the equations can be solved through regions of reversed flow with a prescribed external velocity distribution without meeting any singular behaviour near the separation point (Goldstein singularity).
Next, while the steady boundary layer equations do not permit solutions with $u = \partial u/\partial y = 0$, see Danberg and Fansler (1975), the approximation of Navier-Stokes, as used by Inoue, may permit such solutions because pressure variations in y-direction, being related to transverse diffusion, are allowed.

4.5 Unfolding model for moving wall separations

In this section we continue the discussion on flow patterns along moving walls. A comparison is made between the MRS-model and the unfolding model, as obtained by unfolding the third-order saddle point (section 4.3). Let us start with the downstream moving wall.

Downstream moving wall
Concerning a downstream moving wall Inoue's numerical solution yields flow patterns very similar to those obtained by unfolding the third order saddle point (unfolding model). In the unfolding model, a closed recirculating bubble is formed above the wall when separation is already present at zero wall velocity, see Figure 3.10.
The bubble is bounded by a single streamline which forms a saddle loop with the saddle point (S_3) underneath the bubble. Two stagnation points appear in the flow, the saddle point S_3 just mentioned and a center point C in the interior of the bubble region, see Figure 3.13.
Looking for MRS points in the unfolding model we find two of them, one upstream and one downstream of the bubble; neither of them coincides with S_3, as suggested by the MRS-model. These observations confirm a flow pattern as suggested by the numerical solution given by Inoue. Let us calculate the MRS-points ($u = \partial u/\partial y = 0$) from the unfolding model.

The leading terms describing the streamline pattern are (Eq. (3.21))

$$\dot{x} = u = \mu_o + \mu_1 y + \mu_2 xy + y^2 + \hat{a}_4 x^2 y \qquad (3.23)$$

$$\dot{y} = v = -\frac{1}{2}\mu_2 y^2 - \hat{a}_4 xy^2, \quad \hat{a}_4 > 0$$

and the unfolding model predicts MRS-points in the upper half plane at

$$\text{MRS: } \left(-\frac{\mu_2}{2\hat{a}_4} \pm \frac{1}{2\hat{a}_4}\sqrt{\mu_c - 8\hat{a}_4\sqrt{\mu_o}},\ \sqrt{\mu_o}\right) \qquad (3.24)$$

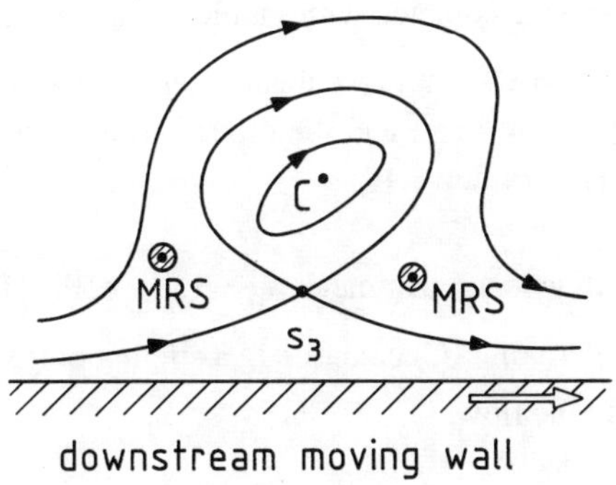

Fig. 3.13. Flow near downstream moving wall according to unfolding model.

Based on the unfolding model the following remarks on the existence and significance of MRS-points are made:

- Real solutions for MRS-points occur only if $\mu_o > 0$, indicating that MRS-points may appear in flows along downstream moving walls; they will not appear in flows along an upstream moving wall.

- MRS-points will appear in the downstream case if the wall velocity parameter, μ_o, does not exceed the critical value $\mu_{o_{crit.}} = (\mu_c/8\,\hat{a}_4)^2$. This critical value equals the value of μ_o where and above which no recirculating region could be observed in the unfolding model. If the bubble disappears due to an increase in wall velocity the MRS-points will vanish at the same time.
- In agreement with Inoue's observation, the unfolding model shows the existence of two MRS-points, one downstream and one upstream of the bubble. However, neither of them will coincide with

$$S_3 : \left(-\frac{\mu_2}{2\hat{a}_4} \, , \, \frac{\mu_c - \sqrt{\mu_c^2 - \mu_o(8\hat{a}_4)^2}}{8\hat{a}_4} \right)$$

as was suggested by the MRS-model.

Only in the two limiting cases, namely zero wall velocity ($\mu_o = 0$) and critical wall velocity ($\mu_o = \mu_{o_{crit.}}$), MRS-points coincide with saddle points in the flow pattern. For zero wall velocity the two MRS-points are located on the wall surface, one of them coincides with the separation point and the other with the attachment point (points S_1 and S_2 respectively in Figure 3.10).

At the critical value of the wall velocity parameter $\mu_o = \mu_{o_{crit.}}$ both MRS-points, the saddle point (S_3) and the center point (C) coalesce in a single point above the wall forming a degenerate point in the flow.

- Except in some particular situations we observe that MRS-points cannot be identified with stagnation points in the flow field signalling the onset of separation. However, MRS-points may be considered as signposts (precursors) indicating the existence of domains with recirculating flow further downstream.

Upstream moving wall

The unfolding model predicts flow patterns near upstream moving walls rather different from those arising in the MRS-context. The MRS-model shows a saddle point (MRS-points) above the wall (Figure 3.12) and a reversed flow region close to the wall. The unfolding model gives a flow pattern with a center point above the wall. This center point is part of a recirculating flow which extends far in streamwise direction forming a layer underneath the main flow. The center of rotation inside the layer is marked by a stagnation point (center point), as shown in Figure 3.14. Moreover, no MRS-points are found in the

unfolding model.
The position of the vortex center (y_c) relative to the maximum height of the layer (y_l), will be determined for the flow pattern defined by Eq. (3.23).
The height of the layer is found from the dividing streamline between main flow and the recirculating region.

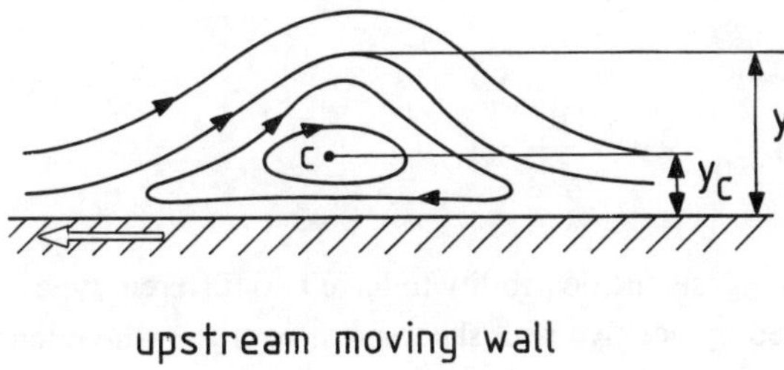

Fig. 3.14. Flow near upstream moving wall according to the unfolding model

The maximum height y_l appears at the vortex position.
The limiting value of the height ratio $h = y_c/y_l$ for zero wall velocity ($\mu_o \uparrow 0$) is found to depend on the sign of μ_c:

$h = 2/3 \qquad \text{if } \mu_c > 0$

$h = 1/3\sqrt{3} \qquad \text{if } \mu_c = 0$

$h = 1/2 \qquad \text{if } \mu_c < 0$

The differences are related to the limiting fixed wall solutions, namely either bubble type separation ($\mu_c > 0$) degenerate flow ($\mu_c = 0$) or fully attached flow ($\mu_c < 0$). Evidently an upstream moving wall interferes more strongly with separated flows than with attached flows.

Finally, we consider the velocity profiles $u = u(y)$ along the wall.
Typical profiles, as obtained from numerical calculations (Inoue, 1981) are shown in Figure 3.15 for varying wall shear stress. In the unfolding model the u-velocity near the wall may be approximated by the leading terms

$$u(x,y) = \mu_o + [\hat{a}_4 x^2 - \frac{\mu_c}{4\hat{a}_4}]\, y + y^2 \qquad (\hat{a}_4 > 0)$$

where x has to be measured with respect to the vortex center. According to this equation, three typical velocity distributions can be observed in recirculating layers, each one corresponding to an x-position where the wall shear stress $\tau(x, o)$ is either negative, zero or positive.

$$\frac{\tau(x,o)}{\mu} = [\frac{\partial u}{\partial y}]_{y=0} = \hat{a}_4 x^2 - \frac{\mu_c}{4\hat{a}_4}$$

The unfolding model suggests the possibility to have two different types of recirculating layers; one characterized by positive wall shear stress over the whole length of the layer ($\mu_c < 0$) and a second type in which the wall shear stress changes sign.

Comparing the results we find a qualitative agreement of the unfolding model with Inoue's calculations.

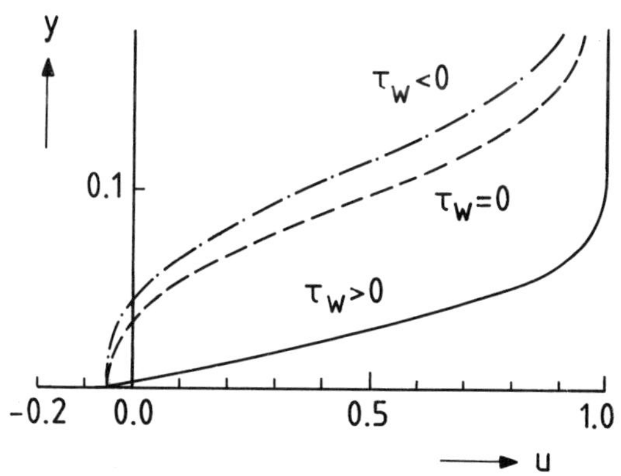

Fig. 3.15. Typical velocity profiles for the case of an upstream moving wall according to Inoue (1981).

5. Unfolding of the topological saddle point of the fifth order

5.1 Description of the unfolding

In this paragraph we will study the topological saddle point of the fifth-order; in these flows separation or attachment occurs tangentially along a wall. In such a separation (attachment) point $\tau = \tau_x = \tau_{xx} = 0$, $\tau_{xxx} \neq 0$ and the shear stress behaves approximately as a cubic function near the point. The streamline pattern near such a point can be obtained from Eq. (3.13) by taking $\hat{a}_4 = 0$, resulting in a system given by the following expression up till third-order terms

$$\begin{aligned} \dot{x} &= y^{-1}.\,u = y + \hat{a}_5 xy + \hat{a}_7 x^3 + \hat{a}_8 x^2 y - \frac{1}{2}\hat{a}_7 xy^2 - \frac{1}{3}\hat{a}_8 y^3 \\ \dot{y} &= y^{-1}.\,v = -\frac{1}{3}\hat{a}_5 y^2 - \frac{3}{2}\hat{a}_7 x^2 y - \frac{2}{3}\hat{a}_8 xy^2 + \frac{1}{2}\hat{a}_7 y^3 \end{aligned} \tag{3.25}$$

Application of Andronov's classification scheme (Chapter I) shows that, if $\hat{a}_7 \neq 0$, the streamline topology near the degenerate point (0,0) is similar to that of the equivalent system

$$\begin{aligned} \dot{x} &= y + \hat{a}_7 x^3 \\ \dot{y} &= -\frac{3}{2}\hat{a}_7 x^2 y \end{aligned} \tag{3.26}$$

The degenerate singularity as given by Eq. (3.26) describes an attaching flow pattern if $\hat{a}_7 > 0$ and a separating flow pattern for $\hat{a}_7 < 0$. In both cases the separating streamline is either approaching or leaving the wall tangentially (Figure 3.16).

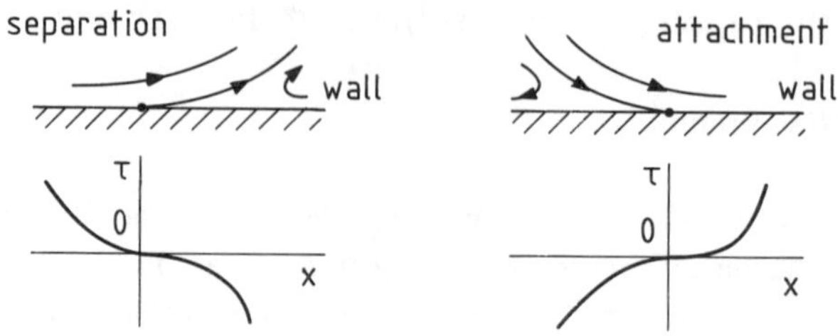

Fig. 3.16. Flow near a fifth-order saddle point.

The wall shear stress distribution near the separation (attachment) point is given by $\tau(x,o) = \hat{a}_7 x^3 + o(x^3)$.

Let us investigate the physical unfolding of Eq. (3.26) satisfying the flow equations together with the no-slip boundary conditions on a fixed wall.
The discussion will be restricted to the case of separation, where $\hat{a}_7 < 0$ (say $\hat{a}_7 = -1$).
The physical unfolding of Eq. (3.26) is obtained if this equation is supplemented with the lower-order terms occurring in Eq. (3.10); the coefficients of these terms serve as the bifurcation parameters. The physical unfolding of Eq. (3.26) becomes

$$\begin{aligned} \dot{x} &= \mu_1 + \mu_2 x + y + \mu_3 x^2 - x^3 = P(x,y) \\ \dot{y} &= -\frac{1}{2}\mu_2 y - \mu_3 xy + \frac{3}{2}x^2 y = Q(x,y) \end{aligned} \tag{3.27}$$

The unfolding, Eq. (3.27), contains three bifurcation parameters, μ_1, μ_2 and μ_3 whose effect on the flow topology have to be studied. Characteristic for the unfolded streamline patterns are number, type and location of the singular points near (0,0). There are two sets of such neighbouring points. First we have the singular points on the wall surfaces (on-wall points) located at

$$\mu_1 + \mu_2 x + \mu_3 x^2 - x^3 = 0, \quad y = 0$$

The second set describes the off-wall singularities given by

$$\frac{1}{2}\mu_2 + \mu_3 x - \frac{3}{2}x^2 = 0, \quad y = -\mu_1 - x(\mu_2 + \mu_3 x - x^2)$$

The on-wall singular points follow from a cubic equation, indicating that one, two or three singularities can occur on the wall. If two singularities appear, the bifurcation is nongeneric since one of the singularities is nonhyperbolic. Nonhyperbolicity of a singularity on the wall can also occur in the case that three singularities coincide. This occurs if the bifurcation parameters obey the relations: $\mu_2 = 3(\mu_3/3)^2$, $\mu_1 = (\mu_3/3)^3$. Then it turns out that no off-wall singularities will arise and the only effect of the bifurcation is a displacement of the fifth-order saddle point along the wall to the location $x = \frac{1}{3}\mu_3$. This nongeneric bifurcation is of no interest and can be ignored without loosing essential features of the unfolding by taking $\mu_3 = 0$ throughout.

This may be explained as follows. The nongeneric perturbation $\mu_2 = 3(\mu_3/3)^2$, $\mu_1 = (\mu_3/3)^3$ occurs in the three-dimensional space (μ_1, μ_2, μ_3) on a spatial curve which approaches the origin (0,0,0) in the direction of the μ_3-axis. The nongeneric subspace has dimension one so that the universal unfolding Eq. 3.27 is fully determined by the two parameters μ_1 and μ_2 having a span which intersects the nongeneric subspace transversally at the origin. Then the on-wall singularities are given by

$$A_i: \ \mu_1 + \mu_2 x_i - x_i^3 = 0, \ y_i = 0, \quad i = 1,2,3$$

whereas the off-wall singularities satisfy

$$B_1, B_2: \ [\pm [\frac{\mu_2}{3}]^{\frac{1}{2}}, -\mu_1 \mp 2[\frac{\mu_2}{3}]^{\frac{3}{2}}], \ \mu_2 > 0$$

The on-wall singularities A_i are elementary or third-order saddle points while the off-wall singular points B_1 and B_2 appear to be a saddle point and a center point, respectively, or a cusp if they coincide.
Nongeneric bifurcations of this unfolding occur on the corresponding bifurcation sets:

$$B_s^{\pm}: \quad \frac{\mu_1}{2} = \pm\left[\frac{\mu_2}{3}\right]^{\frac{3}{2}}, \quad B_c: \quad \mu_2 = 0 \; (\mu_1 < 0)$$

These sets, together with the various resulting flow patterns, are shown in Figure 3.17. The generic bifurcation of a fifth-order saddle point will give one of the following three topologically different structurally stable flow patterns.

The first and simplest case consists of a single hyperbolic saddle point at the wall, corresponding to the ordinary separation case already described as the classical Oswatitsch-Legendre solution. It occurs if $\mu_2 < 0$ and if $\mu_2 > 0$, $\frac{\mu_1}{2} > \left[\frac{\mu_2}{3}\right]^{\frac{3}{2}}$.

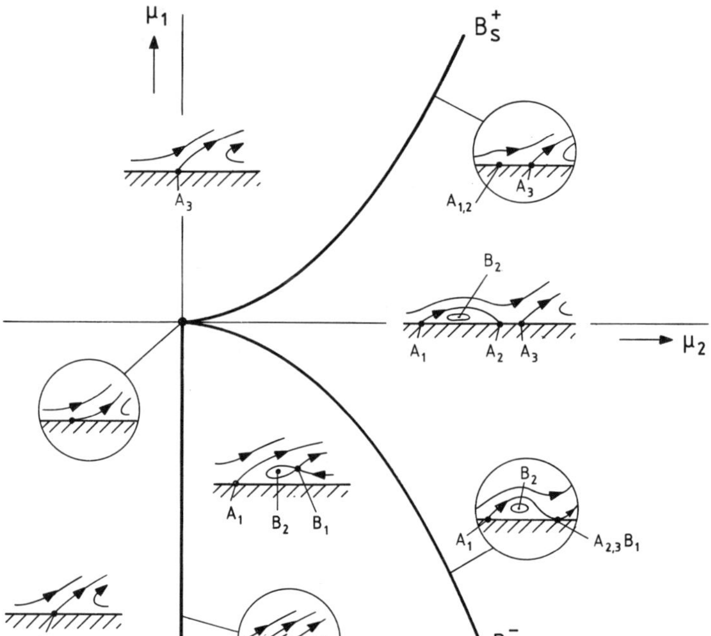

Fig. 3.17. Unfolding of fifth order saddle point, bifurcation sets:

$$B_s^{\pm}: \quad \frac{\mu_1}{2} = \pm\left[\frac{\mu_2}{3}\right]^{\frac{3}{2}}, \quad B_c: \quad \mu_2 = 0 \; (\mu_1 < 0).$$

The second flow pattern looks less familiar; it contains a well-known separation point on the wall but in addition a center point and a saddle point in the flow downstream of the separation point. The upstream separatrices of the saddle point form a streamline which encircles the center point by a closed loop. The flow region behind the separation streamline contains a closed recirculating region off the wall surface.Finally, the third structurally stable flow pattern shows the familiar laminar separation bubble followed by a secondary separation further downstream.

It occurs for $\mu_2 > 0$, $-[\frac{\mu_2}{3}]^{\frac{3}{2}} < \frac{\mu_1}{2} < [\frac{\mu_2}{3}]^{\frac{3}{2}}$.

Transitional flow patterns between these three types of flow are described by the nongeneric bifurcations. They correspond to structurally unstable flow patterns shown in Figure 3.17. Again, we distinguish three different possibilities:

1. B_c: In the flow region downstream of the separation streamline a nonhyperbolic cusp singularity occurs. The cusp singularity may bifurcate in two ways: it will fall apart into a center point and a saddle point, or it disappears leaving no singularity in its neighbourhood.
2. B_s^- : A third-order saddle point with three hyperbolic sectors occurs on the wall surface. Bifurcation of the third-order saddle point (see paragraph 4.1) results either in two hyperbolic saddles on the wall or in a single saddle point above the wall.
3. B_s^+ : A third-order saddle point with one hyperbolic sector occurs on the wall surface. Bifurcation gives either a laminar separation bubble on the wall, or it disappears leaving no singularity in its neighbourhood. Note that this bifurcation process has already been described in paragraph 4 of this Chapter.

5.2 Bubble capturing by a secondary separation

It will be attempted now to interpret the unfolding of the fifth-order topological saddle point in a practical flow situation where a separation bubble is followed by a secondary separation. For this purpose it is convenient to observe the various flow patterns if μ_2 varies at constant μ_1. It suffices to take $\mu_1 < 0$ since all possible flow patterns will then be encountered and the variation of μ_2 may be restricted to go from an appropriate positive value $\mu_2 > 3[-\frac{\mu_1}{2}]^{\frac{2}{3}}$ to $\mu_2 < 0$.

These flow patterns may be thought to be embedded in the flow field over an airfoil, more in particular they would appear in the process of the capture of a laminar separation bubble by a secondary separation, Figure 3.18.

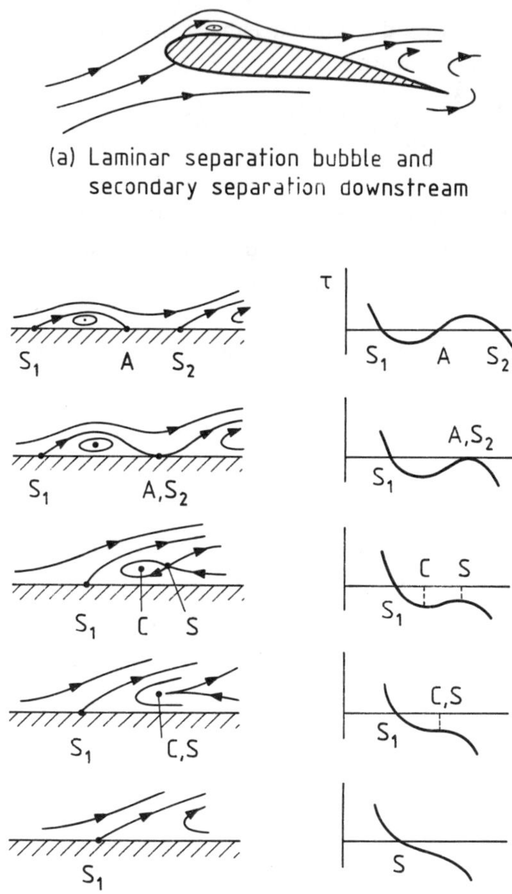

Fig. 3.18. Bubble capturing by upstream moving separation.

For $\mu_2 > 3[-\frac{\mu_1}{2}]^{\frac{2}{3}}$ we have the situation as given in Figure 3.18a. If μ_2 decreases below this value, the secondary separation points moves in upstream direction, approaching the laminar separation bubble. At $\mu_2=3[-\frac{\mu_1}{2}]^{\frac{2}{3}}$ the secondary separation starts to interfere with the bubble: the question arises how the bubble will be 'captured' by the upstream moving separation.

The unfolding of the fifth-order topological saddle point suggests (Figure 3.18b) the following answer: before a final situation is reached in which the bubble is completely lost in the separated flow region ($\mu_2 < 0$), there appears a transitional stage with an off-wall recirculating region downstream of the separation streamline. The appearance of such a transition stage, which is structurally stable, is reflected very well by the wall shear stress distribution, Figure 3.18b.

When the secondary separation point S_2 meets the rear part of the separation bubble at the attachment point A, the wall shear stress reaches a local maximum value $\tau = 0$.

As long as the recirculating region exists, the wall shear stress:

$$\tau = \mu(\mu_1 + \mu_2 x - x^3 + O(3))$$

shows two extreme values downstream of the separation point S_1; a local minimum and a local maximum.

The minimum corresponds to the x-position of the center of the vortex and the maximum to the x-position of the free saddle point. The recirculating region disappears if the wall shear stress distribution shows an inflexion point where the shear stress gradient vanishes. Finally, it may be noted that the surface pressure gradient remains positive in all these different flow patterns.

6. Unfolding of a saddle point with three hyperbolic sectors in a half plane, $\tau_{xx} \neq 0$

Paragraphs 6 and 7 will be devoted to the classification of unfoldings of degeneracies in the case of zero pressure gradient. Referring to paragraph 3.2 where these degeneracies are described, we can distinguish between the cases $\tau_{xx} \neq 0$ and $\tau_{xx} = 0$. Each case will be considered in a separate paragraph. Let us start with the less complicated one, $\tau_{xx} \neq 0$.

6.1 Universal physical unfolding

The corresponding degenerate flow is governed by the quadratic part of Eq. (3.14) ($a_4 \neq 0$)

$$\dot{x} = a_4x^2 + a_5xy - \frac{2}{3}a_4y^2$$
$$\dot{y} = -a_4xy - \frac{1}{3}a_5y^2 \tag{3.28}$$

The degenerate flow pattern in the semi-plane $y \geq 0$ consists of a saddle point with four separatrices which divide the flow into three hyperbolic sectors, Figure 3.19.

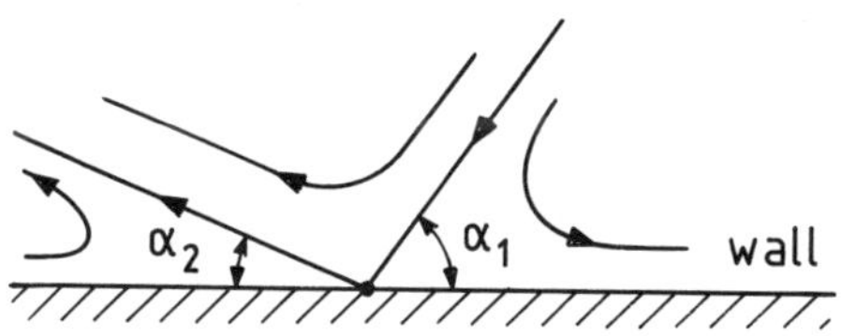

Fig. 3.19. Degenerate flow at $p_x=0$, $\tau_{xx}\neq 0$.

The physical unfolding satisfying the flow equations together with the no-slip boundary condition may be obtained from Eq. (3.10) by taking a_1, a_2 and a_3 as bifurcation parameters, each having zero as the bifurcation value. As usually, we introduce the parameters μ_1, μ_2 and μ_3 to describe small perturbations of a_1, a_2 and a_3 with respect to their bifurcation value. Since $a_4 \neq 0$ we may write

$$\hat{a}_1 = \mu_1,\ \hat{a}_2 = \mu_2,\ \hat{a}_3 = \mu_3$$

where $\hat{a}_i = a_i/a_4$, and the unfolding of Eq. (3.28) becomes

$$\begin{aligned} \dot{x} &= \mu_1 + \mu_2 x + \mu_3 y + x^2 + \hat{a}_5 xy - \frac{2}{3} y^2 + O(3) \\ \dot{y} &= -\frac{1}{2}\mu_2 y - xy - \frac{1}{3}\hat{a}_5 y^2 + O(3) \end{aligned} \tag{3.29}$$

where a_4 is taken in the parameter t.

6.2 Movement Principle

The unfolding contains three bifurcation parameters, but it will be shown that only two of them are really necessary for a universal unfolding; thus the unfolding has codimension two.

If the unfolding has codimension two, the three-dimensional parameter space (μ-space) must contain a nongeneric subspace, which represents parameter combinations for which the original degeneracy (Eq. (3.28)) is preserved after perturbation. This nongeneric subspace represents flow patterns which are topologically equivalent to that of the degenerate point, apart from inconsequential translations in the x,y-frame. Due to the no-slip boundary conditions, it is obvious that only those translations are allowed which will shift the degenerate point along the wall surface.

To find this nongeneric subspace, we will adopt here the socalled *Movement Principle* which describes the *degenerate* flow pattern in a new reference frame $\bar{x},\bar{y}$ with $\bar{x} = x - x_o$, $\bar{y} = y$ so that $(-x_o, o)$ represents the location of the degenerate point in this new reference frame.

Equation (3.28) is then recast into the form

$$\begin{aligned} \dot{\bar{x}} &= x_o^2 + 2x_o\bar{x} + \hat{a}_5 x_o\bar{y} + \bar{x}^2 + \hat{a}_5\bar{x}\bar{y} - \frac{2}{3}\bar{y}^2 \\ \dot{\bar{y}} &= -x_o\bar{y} - \bar{x}\bar{y} - \frac{1}{3}\hat{a}_5\bar{y}^2 \end{aligned} \tag{3.30}$$

Comparing the unfolding of the degenerate point in the original system (Eq. (3.29)) with the representation of the degeneracy in a shifted coordinate system (Eq. (3.30)) it may be concluded that the perturbation which satisfies

$$\mu_1 = k\, x_o^2$$

$$\mu_2 = k\, 2x_o$$

$$\mu_3 = k\, \hat{a}_5 x_o$$

where k is a constant, is nongeneric since it leaves the degenerate flow pattern unaltered apart from a shift along the wall surface.

The nongeneric perturbation occurs in the three-dimensional space (μ_1, μ_2, μ_3) on a spatial curve which approaches the origin (0,0,0) in the direction $(0,2,\hat{a}_5)$, see Figure 3.20.

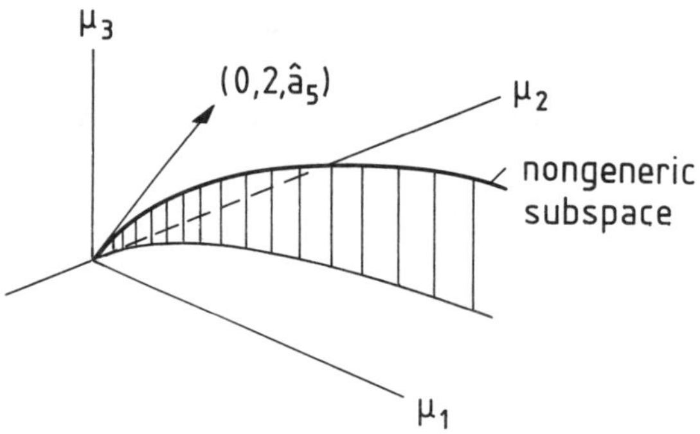

Fig. 3.20. Nongeneric subspace of Eq. 3.29.

Consequently, the corresponding nongeneric subspace has dimension one, implying that a universal unfolding can be fully determined by two parameters if these two parameters are chosen such that their span in the parameter space intersects the nongeneric subspace transversally at the origin.

Since $\hat{a}_5$ is finite the plane $\mu_2 = 0$ satisfies the transversality condition, so that the following two-parameter family provides a universal unfolding of Eq. (3.28):

$$\dot{x} = \mu_1 + \mu_3 y + x^2 + \hat{a}_5 xy - \frac{2}{3} y^2 + O(3)$$

$$\dot{y} = -xy - \frac{1}{3} \hat{a}_5 y^2 + O(3)$$

With suitable rescaling of x, y and t, all possible cases for $\hat{a}_5 < 0$ can be reduced to $\hat{a}_5 > 0$.

6.3 Bifurcation sets, flow patterns

To find the bifurcation sets for which the flow pattern is structurally unstable we seek the singular points near (0,0) on the wall as well as in the flow field.
The on-wall singularities S_1, S_2 are given by $y = 0, x = \pm\sqrt{-\mu_1}$ and will occur only if $\mu_1 < 0$.
The off-wall singularities satisfy

$$C_{1,2}: \quad x_c = -\frac{\hat{a}_5}{3} y_c, \quad y_c = +\frac{9\mu_3}{4K^2} \mp \frac{9}{4K^2}\sqrt{\mu_3^2 + \frac{8}{9}K^2\mu_1}$$

where $K = \sqrt{(3 + \hat{a}_5^2)}$.
For the generic bifurcations, the on-wall singularities S_1, S_2 are saddle points, while the off-wall singularities C_1, C_2 appear to be a center point (C_1) and a saddle point (C_2) respectively.

The nongeneric bifurcations are specified by $\mu_1 = 0$, and $\mu_3^2 + \frac{8}{9}K^2\mu_1 = 0$ and will be denoted as B_s and B_c respectively. This bifurcation set, together with the various flow patterns in different regions (I-V) are presented in Figure 3.21. At $\mu_1 = 0$ and $\mu_3 > 0$ we have a third-order saddle point at the wall (where the points S_1, S_2 and C_1 coalesce) and a single saddle point (C_2) above the wall.

Small perturbations due to μ_1 variations with respect to $\mu_1 = 0$ ($\mu_3 \neq 0$) give third-order saddle point bifurcation as described earlier in paragraph 4. If μ_1 is negative and $\mu_3 > 0$ a flow pattern results with a separation bubble at the wall underneath the saddle point C_2 (region IV); if μ_1 is taken as positive, only the saddle point C_2 remains in the flow above the wall (region V).

At B_c, cusp point bifurcation occurs in the flow, causing either the formation of a recirculating bubble (region III) in the flow domain in between a separation and an attachment streamline if μ_3 is increased or a disappearance of the cusp point if μ_3 decreases (region II).

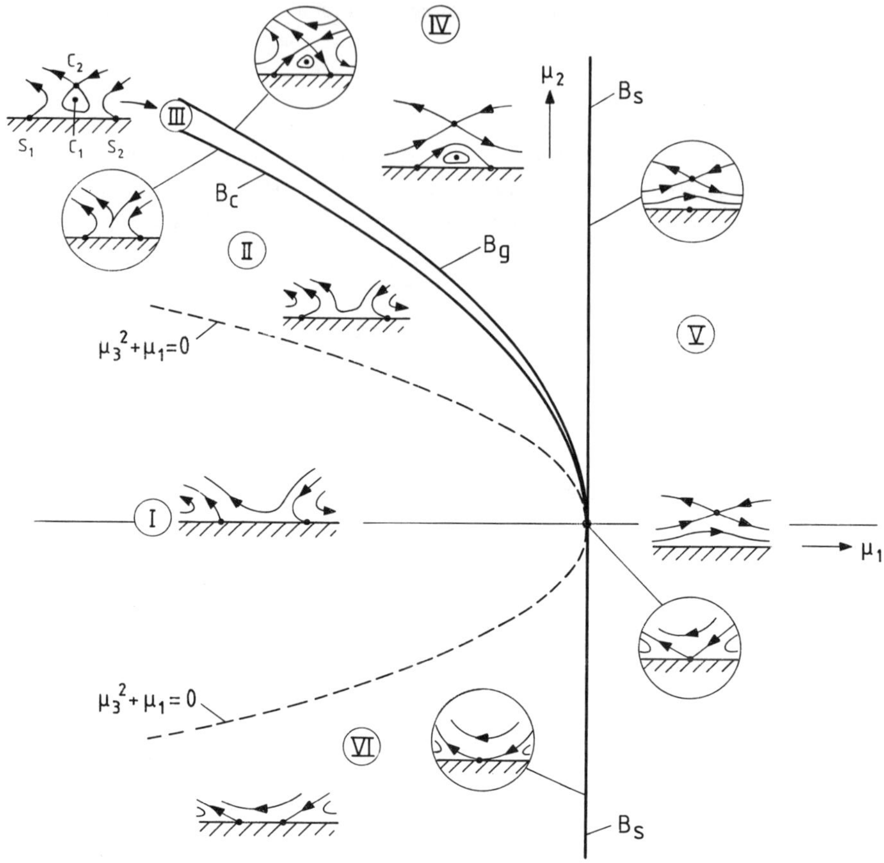

Fig. 3.21. Unfolding of saddle point with three hyperbolic sectors, bifurcation sets: B_s*:* $\mu_1 = 0$, B_c*:* $\mu_3 = \frac{2}{3}\sqrt{-2\mu_1}$, $B_g : \mu_3 = K\sqrt{-\mu_1}$.

We now note that the phase portraits in region III near B_c and in region IV near $\mu_1 = 0$ ($\mu_3 > 0$) are not homeomorphic, for the flow in IV possesses a heteroclinic cycle whereas the flow in III has a homoclinic cycle.

Hence there must be an additional bifurcation which transforms the phase portraits in III and IV into each other. The saddle C_2 and the center C_1 still exist in the region III and IV and thus a *global* bifurcation must take place, perhaps a saddle-point-triangle formed by saddle-saddle connections (ss) between the saddle points S_1, S_2 and C_2 and surrounding the center point C_1.

To study this global bifurcation, shown in Figure 3.22, we determine the Hamiltonian of

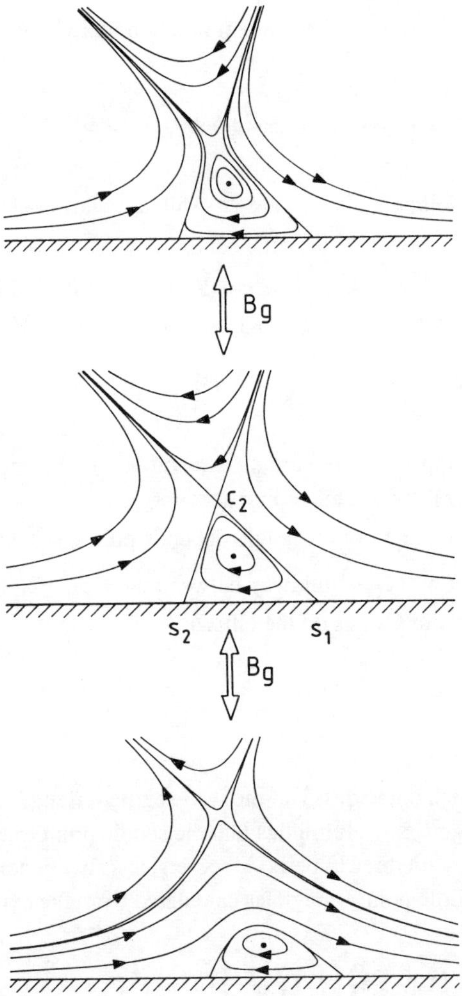

Fig. 3.22. Global bifurcation of saddle point triangle.

the vector field:

$$\dot{x} = \mu_1 + \mu_3 y + x^2 + \hat{a}_5 xy - \frac{2}{3} y^2 + O(3)$$
$$\dot{y} = -xy - \frac{1}{3} \hat{a}_5 y^2 + O(3)$$

which gives the leading terms of the stream function $\psi(x,y)$ near (0,0):

$$\psi(x,y) = \frac{y^2}{6} [3\mu_1 + 2\mu_3 y + 3x^2 + 2\hat{a}_5 xy - y^2 + O(3)] \qquad (3.31)$$

Because the saddle points S_1 and S_2 are on-wall singularities, the presumed ss-connections S_1-C_2 and S_2-C_2 have to satisfy the condition $\psi(x,y) = 0$.
A simple calculation shows that these conditions are fulfilled if $\mu_3 = K \sqrt{-\mu_1}$, then a saddle-point-triangle is formed by the saddle points at $(-\sqrt{-\mu_1},0)$, $(\sqrt{-\mu_1},0)$, and $\frac{\sqrt{-\mu_1}}{K}(-\hat{a}_5,3)$ and a center point in $\frac{\sqrt{-\mu_1}}{2K}(-\hat{a}_5,3)$.

The ss-connections appear to be straight streamlines if the higher-order terms in Eq. (3.31) denoted by O(3), are not taken into account.
The geometry of the saddle point triangle depends on the scaled variable $\hat{a}_5$, i.e. it depends on the ration of $p_{xx}(\tau_{xx})^{-1}$ midway between the saddle points on the wall.
The off-wall saddle point C_2 lies on the ellipse:

$$3[\frac{x}{l}]^2 + [\frac{y}{l}]^2 = 3$$

the minor axis of which coincides with the wall segment (length $2l$) between the on-wall saddle points, see Figure 3.23. It implies that the saddle point triangle, as observed in the unfolding of a saddle with three hyperbolic sectors, has always acute angles. The vorticity distribution in the saddle point triangle as calculated from the expression

$$\omega = l^2\{1 - 2K[\frac{y}{l}] - 2\hat{a}_5[\frac{x}{l}][\frac{y}{l}] - [\frac{y}{l}]^2 - [\frac{x}{l}]^2\}$$

is shown in Figure 3.23 as well.

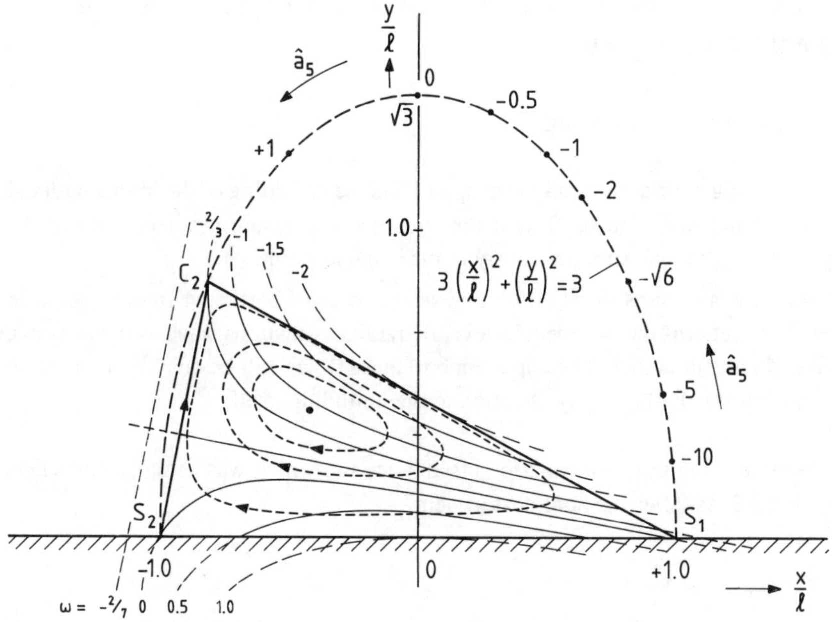

Fig. 3.23. Location of off-wall saddle point and vorticity contours for
$\hat{a}_5 = p_{xx}/\tau_{xx} = \sqrt{6}$.

7. Unfolding of a saddle point with two or four hyperbolic sectors in a half plane, $\tau_{xx} = 0$

7.1 Universal physical unfolding

The last unfolding we want to study in chapter III is the unfolding of the higher-order singularity occurring in a viscous flow if the streamwise pressure gradient and the shear stress gradient τ_{xx} vanish simultaneously in the singular point.

This singularity, and parts of its unfolding, seems to be of particular interest because it describes flow patterns that we meet in several practical situations, such as the flow in the near wake of smooth bodies. The application to these flows will be dealt with in the next paragraphs, but we shall first pay attention to the unfolding itself.

The degenerate singularity we want to unfold here, and which was actually investigated in paragraph 3.2, satisfies the nonlinear system

$$\begin{aligned} \dot{x} &= a_5 xy + a_7 x^3 \\ \dot{y} &= -\frac{1}{3} a_5 y^2 - \frac{3}{2} a_7 x^2 y, \qquad a_5 \neq 0,\ a_7 \neq 0 \end{aligned} \tag{3.32}$$

The corresponding phase portraits in the semiplane $y \geq 0$, see Figure 3.8, showed that four different flow patterns will arise for various choices of a_5 and a_7.

A separate treatment of all these four flow patterns is not necessary because Eq. (3.32) is invariant for the transformations $(a_5,y) \rightarrow (-a_5,-y)$ and $(a_7,y,t) \rightarrow (-a_7,-y,-t)$.

Thus if Eq. (3.32) is studied both in the upper ($y \geq 0$) as well as in the lower ($y \leq 0$) half plane, it is sufficient to discuss the possible unfoldings of Eq. (3.32) for at most one case, say $a_5 < 0$ and $a_7 < 0$. The corresponding higher-order singularity is shown in Figure 3.24, it represents flow separation where the pressure attains a maximum at the wall; suitable rescaling allows to take $a_5 = a_7 = -1$.

The shearstress is a cubic function near the singularity, it obeys the relation $\tau = \mu\,(a_7\,x^3 + O(4))$

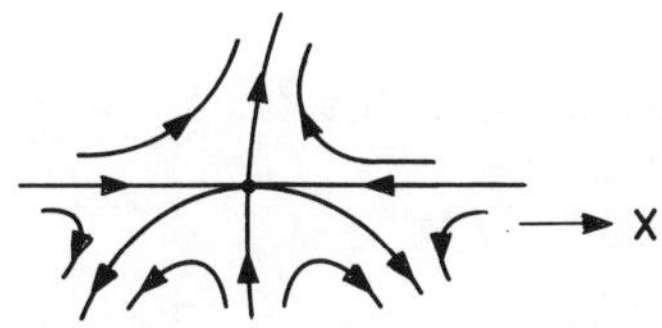

Fig. 3.24. Degenerate flow at $p_x = 0$, $\tau_{xx} = 0$.

Similarly to former cases, the physical unfolding may be obtained from Eq. (3.10) by taking the 'lost' parameters (a_1,a_2,a_3,a_4) as the bifurcation parameters.
All of them have zero as the bifurcation value, and as usual the parameters μ_1, μ_2, μ_3 and μ_4 will be introduced to denote small perturbations of a_i (i = 1 to 4). Then we find the following physical unfolding:

$$\begin{aligned}\dot{x} &= \mu_1 + \mu_2 x + \mu_3 y + \mu_4 x^2 - xy - \frac{2}{3}\,\mu_4 y^2 - x^3 + O(3)\\ \dot{y} &= -\frac{1}{2}\,\mu_2 y - \mu_4 xy + \frac{y^3}{3} + \frac{3}{2}\,x^2 y + O(3)\end{aligned} \qquad (3.33)$$

7.2 Determination of codimension

The number of relevant parameters determining the unfolding completely can be reduced if the four-dimensional parameter space M: $(\mu_1,\mu_2,\mu_3,\mu_4)$ contains a nongeneric subspace where the original degeneracy (Eq. (3.32)) is preserved after perturbation.
Such a situation will occur if the effect of the perturbations is limited to a shift of the degenerate point along the wall surface. Other possibilities which preserve the degenerate point, such as a rotation in the x,y-plane or a shift off the wall, must be ruled out for their violation of the no-slip boundary conditions.
Let us elaborate the shift along the wall by using the movement principle (par. 6.2) and write Eq. (3.32) with $a_5 = a_7 = -1$ in the translated coordinate system $\bar{x} = x - x_0$, $\bar{y} = y$

$$\dot{\bar{x}} = -x_o^3 - 3x_o^2\bar{x} - x_o y - 3x_o\bar{x}^2 - \bar{x}\bar{y} - \bar{x}^3$$
$$\dot{\bar{y}} = \frac{3}{2}x_o^2\bar{y} + 3x_o\bar{x}\bar{y} + \frac{\bar{y}^2}{3} + \frac{3}{2}\bar{y}\bar{x}^2 \tag{3.34}$$

Comparing this result with the unfolding (Eq. (3.33)) we observe that the particular perturbation

$$\mu_1 = -kx_o^3$$
$$\mu_2 = -3kx_o^2$$
$$\mu_3 = -kx_o$$
$$\mu_4 = -3x_o$$

where k is a constant, is the nongeneric perturbation which leaves the degenerate phase portrait unaffected.

This nongeneric perturbation is a one-dimensional subspace of M and approaches the bifurcation point (0,0,0,0) in the nongeneric direction (0,0,-1,-3). The unfolding of Eq. (3.32) is fully determined by three parameters, which have to be chosen in such a way that their span in M actually intersects the nongeneric subspace transversally in the bifurcation point. With regard to the nongeneric direction, the transversality condition will be satisfied throughout by taking $\mu_4 = 0$ so that the following three-parameter family provides a *codimension-three* unfolding of Eq. (3.32).

$$\dot{x} = \mu_1 + \mu_2 x + \mu_3 y - xy - x^3 + O(3)$$
$$\dot{y} = -\frac{1}{2}\mu_2 y + \frac{1}{3}y^2 + \frac{3}{2}x^2 y + O(3) \tag{3.35}$$

7.3 Neighbouring singular points, local bifurcation sets B_s and B_c

To obtain insight into the various properties of the phase portraits of Eq. (3.35) we seek the singular points, $\underline{x}_i = (x_i, y_i)$, near (0,0) on the wall as well as in the flow field. Values of the parameters for which these singular points are nonhyperbolic are the bifurcation sets at which the flow patterns are structurally unstable. Since we only deal with phase portraits close to the origin (x,y)= (0,0) we are only interested in the local geometry of the surfaces $x_i = x_i(\mu_1,\mu_2,\mu_3)$ and $y_i = y_i(\mu_1,\mu_2,\mu_3)$ near the bifurcation point $\mu_1 = \mu_2 = \mu_3 = 0$ where $x_i = 0$.

The parametrisation $x_i = \bar{x}_i \mu_1^{\frac{1}{3}}$, $y_i = \bar{y}_i (\mu_1)^{\frac{2}{3}}$, $\mu_2 = k_2 \mu_1^{\frac{2}{3}}$, $\mu_3 = k_3 \mu_1^{\frac{1}{3}}$, allows to find these surfaces and phase portraits from the truncated system

$$\begin{aligned} \dot{x} &= \mu_1 + \mu_2 x + \mu_3 y - xy - x^3 \\ \dot{y} &= -\frac{1}{2}\mu_2 y + \frac{1}{3} y^2 + \frac{3}{2} x^2 y \end{aligned} \tag{3.36}$$

The *on-wall singularities* have x-locations on the wall found as solutions of the cubic equation $\mu_1 + \mu_2 x - x^3 = 0$.
This equation points out that the number of on-wall singularities depends on μ_1 and μ_2; actually three singularities appear if $(\mu_2/3)^3 > (\mu_1/2)^2$ whereas one singular point may be found if $(\mu_2/3)^3 < (\mu_1/2)^2$. If $(\mu_2/3)^3 \neq (\mu_1/2)^2$ it may be verified that the singular points are saddle points and that all of them are hyperbolic. The condition

$$B_s: \quad (\mu_2/3)^3 = (\mu_1/2)^2$$

forms a bifurcation set, called B_s.
At the bifurcation set B_s, there will be at most two on-wall singularities of which at least one is nonhyperbolic. The type of the nonhyperbolic point is established as follows.
Let us denote the x-position of the nonhyperbolic point by x_1 and that of the remaining saddle point by x_2. Since x_1 is at least a double root of the cubic equation $\mu_1 + \mu_2 x - x^3 = 0$ we find easily that $x_2 = -2x_1$ and $x_1 = -(\mu_1/2)^{1/3}$.
If $(x_1,0)$ is taken as a new origin in the phase plane, system (3.36) becomes

$$\begin{aligned} \dot{x} &= [\mu_3 + [\frac{\mu_1}{2}]^{1/3}]\, y + 3\,[\frac{\mu_1}{2}]^{1/3} (x-x_1)^2 - (x-x_1)\, y - (x-x_1)^3 \\ \dot{y} &= -3\,[\frac{\mu_1}{2}]^{1/3} (x-x_1)\, y + \frac{1}{3} y^2 + \frac{3}{2} (x-x_1)^2\, y \end{aligned} \tag{3.37}$$

Unless $(\mu_2,\mu_3) = [3(\mu_1/2)^{2/3}, -(\mu_1/2)^{1/3}]$, we can use Andronov's theorems 66 and 67 to establish that $(x_1,0)$ is a *third-order saddle point* if $\mu_1 \neq 0$ or a *fifth-order saddle point* if $\mu_1 = 0$.

The particular case: $(\mu_2,\mu_3) = [3(\mu_1/2)^{2/3}, -(\mu_1/2)^{1/3}]$, being element of B_s will be labelled B_1. It describes a nonhyperbolic saddle point with three separatrices having the angular coefficients 0, $-\frac{9}{2}(\mu_1/2)^{1/3}$ and ∞.

Finally we conclude that the bifurcation set $B_s \backslash B_1$ indicates saddle point bifurcations on the wall; of third order for $\mu_1 \neq 0$ and of fifth order for $\mu_1 = 0$. Both types of bifurcation have already been discussed (paragraph 4 and 5) so that the results can be applied here without further comments.

The positions of *off-wall singular points* of Eq. (3.36) follow from the relations

$$7x^3 - 9\mu_3 x^2 - \mu_2 x + 2\mu_1 + 3\,\mu_2\mu_3 = 0, \quad y = \frac{3}{2}(\mu_2 - 3x^2) \tag{3.38}$$

The x-component satisfies a cubic equation, with the result that if $(7^2/3^3\,\mu_1 + 7/3\mu_2\mu_3 - \mu_3^3)^2 \neq (\mu_3^2 + 7/3^3\,\mu_2)^3$ one or three hyperbolic points appear. The condition

$$B_c: \quad (\frac{7^2}{3^3}\mu_1 + \frac{7}{3}\mu_2\mu_3 - \mu_3^3)^2 = (\mu_3^2 + \frac{7}{3^3}\mu_2)^3$$

forms a bifurcation set indicating the existence of nonhyperbolic points in the flow field. Let us examine the topological type of this nonhyperbolic point in order to determine the character of B_c.

If (x_1,y_1) is a double root of Eq. (3.38), x_1 satisfies the relation $21x_1^2 - 18\mu_3 x_1 - \mu_2 = 0$ and the linear part of the vector field (Eq. (3.36)):

$$Df = \frac{1}{3}\begin{pmatrix} -y & 3(\mu_3 - x) \\ 9xy & y \end{pmatrix}$$

has zero as a double eigenvalue in (x_1,y_1).

The linear transformation

$$T = \frac{1}{3}\begin{pmatrix} 3 & 0 \\ -y_1 & 3(\mu_3 - x_1) \end{pmatrix}$$

which is nonsingular on $B_c \backslash B_1$ allows Eq. (3.36) to be brought into the normal form

$$\dot{\hat{x}} = \hat{y} - \left(\frac{2}{9}\frac{y_1}{\mu_3 - x_1}\right)\hat{x}^2 - \frac{\hat{x}\hat{y}}{\mu_3 - x_1} - \hat{x}^3$$

$$\dot{\hat{y}} = \frac{y_1}{6(\mu_3 - x_1)}(9\mu_3^2 - 12\mu_3 x_1 + \mu_2)\,\hat{x}^2 + \frac{4}{9}\frac{y_1}{\mu_3 - x_1}\hat{x}\hat{y} + \frac{1}{3(\mu_3 - x_1)}\hat{y}^2 + \frac{3}{2}\hat{x}^2\hat{y}^2 + \frac{5}{6}y_1\hat{x}^3$$

where $\hat{x} = x - x_1$ and $\hat{y} = -\frac{1}{3}y_1(x - x_1) + (\mu_3 - x_1)(y - y_1)$.

Using Andronov's theorems 66 and 67 it follows that the nonhyperbolic point (x_1, y_1) is a *cusp point* singularity provided that the $\dot{\hat{y}}$-equation has a non-vanishing finite $\hat{x}^2$ term, thus if $9\mu_3^2 - 12\mu_3 x_1 + \mu_2 \neq 0$.

This condition is fulfilled except for the case $\mu_1 = \mu_2 = 0$ and for the case labelled B_2 represented by the parameter combinations:

$$B_2: \quad [\mu_2, \mu_3] = \left[-3\sqrt[3]{\frac{7}{25}}\left(\frac{\mu_1}{2}\right)^{2/3}, \frac{1}{3}\sqrt[3]{\frac{49}{5}}\left(\frac{\mu_1}{2}\right)^{1/3}\right], \qquad B_2 \subset B_c$$

For $\mu_1 = \mu_2 = 0$ there occurs a *fifth-order saddle point* that coincides with a singularity on the wall whereas B_2 represents a *third-order saddle point* in the half plane below the wall.

In conclusion: the bifurcation set $B_c \backslash (B_1 \cup B_2)$ corresponds to cusp point bifurcation in the flow off the wall. In paragraph 5.1 cusp point bifurcations have already been discussed and we have seen there that they cause the spontaneous generation of singular points in regular domains of flow fields. Since the vector field (Eq. (3.36)) is Hamiltonian, these new singularities will always appear as a center-saddle pair.

7.4 Flow patterns and global bifurcation sets B_{g1} and B_{g2}

The bifurcation sets B_s and B_c divide the parameter space (μ_1,μ_2,μ_3) in several subdomains where flow patterns of different topology may occur. Standard methods, such as parametrization and singular-point analysis (c.f. previous paragraphs) may be used to establish these patterns in either subdomain separately. The task to obtain a clear three-dimensional view of the parameter space and of its partition by B_s and B_c in subdomains is not an easy one; especially the geometry of B_c is rather complicated.

Therefore, we investigate the case $\mu_1 = 0$ first and then the influence of $\mu_1 \neq 0$ will be considered.

For $\mu_1 = 0$ Eq. (3.36) reduces to

$$\dot{x} = \mu_2 x + \mu_3 y - xy - x^3$$
$$\dot{y} = -\frac{1}{2}\mu_2 y + \frac{y^2}{3} + \frac{3}{2}x^2 y \tag{3.39}$$

The bifurcation set B_s: $(\mu_2/3)^3 = (\mu_1/2)^2$ which concerns the on-wall singularities, reduces for $\mu_1 = 0$ to the simple form $\mu_2 = 0$ and coincides with the μ_3-axis. However, if $\mu_1 = 0$ the μ_3-axis also forms part of the B_c bifurcation set. For μ_1=0 the bifurcation set B_c becomes

$$\left(\frac{7}{3}\mu_2\mu_3 - \mu_3^3\right)^2 = \left(\mu_3^2 + \frac{7}{3^3}\mu_2\right)^3$$

which consists of three branches $\mu_2 = 0$, $\mu_3^2 = 0{,}00334\,\mu_2$ and $\mu_3^2 = 0.95963\,\mu_2$.

A sketch of B_s and B_c in the μ_2,μ_3-plane is shown in Figure 3.25. This figure shows also the flow patterns caused by unfolding the degenerate singularity near the bifurcation point $(\mu_2,\mu_3) = (0,0)$. It suffices to take $\mu_3 > 0$ only. Flow patterns corresponding to $\mu_3 < 0$ can be obtained from those at $\mu_3 > 0$ by a mirror reflection with respect to the y-axis.

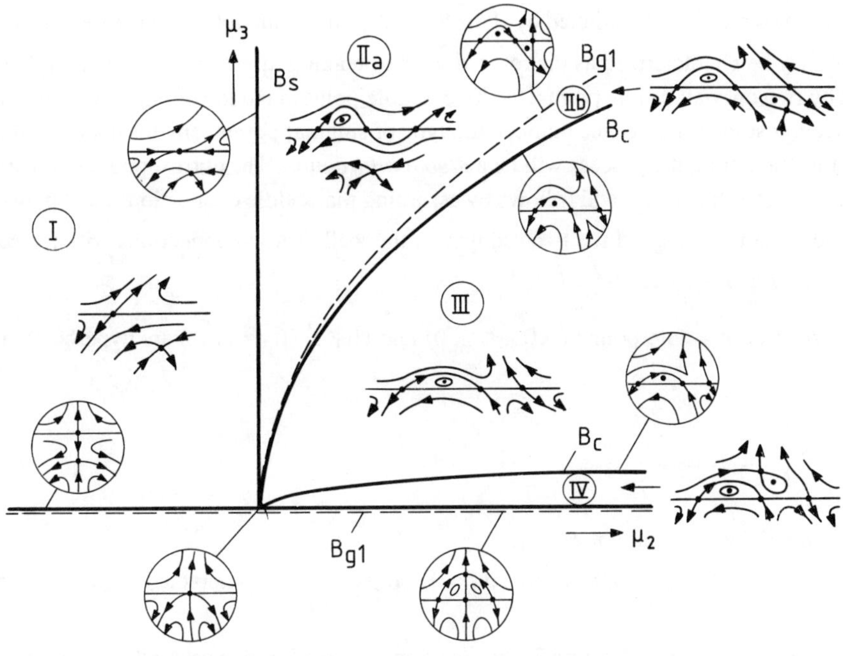

Fig. 3.25. Partial unfolding of Eq. 3.36 ($\mu_1 = 0, \mu_3 > 0$) bifurcation sets:
B_s: $\mu_2=0$, B_c: $\mu_3=0.9796\sqrt{\mu_2}$, $\mu_3=0.05789\sqrt{\mu_2}$, B_{g1}: $\mu_3=0$, $\mu_3=\sqrt{\mu_2}$.

Let us describe the various flow pattern if μ_2 increases at constant $\mu_3 > 0$.

In region I we find two hyperbolic saddle points, one on the wall and the other in the lower semi-plane ($y < 0$). Going to region IIa we approach the line $\mu_2=0$ where the saddle point on the wall becomes a degenerate fifth-order saddle point. Crossing the line $\mu_2 = 0$ will unfold this degenerate saddle point as described in paragraph 5, Figure 3.17, whereas the hyperbolic saddle in the lower half plane remains qualitatively unaffected.

Consequently, the flow region in IIa contains two separation bubbles, one on either side of the wall, and a 'free' saddle point below the wall. In region IIb, we find that the on-wall separation bubble in the lower half plane (present in IIa) does not occur but is replaced by a center-saddle pair below the wall. Obviously there is a transition from IIa to IIb (and vice versa) which is not predicted by B_s or B_c. Detailed calculations show us that this transition may be characterized as the formation and breaking up of a separatrix triangle formed by two on-wall saddle points, the 'free' saddle point in the lower half plane and their connecting streamlines. Since number and type of singular points remains unchanged during this transition, the process will be a *global bifurcation.* The corresponding global bifurcation set is called B_{g1} and follows by assuming that saddle connections exist between the saddle in the flow and the two saddles on the wall. These connections will appear if the condition $\mu_1 + \mu_2\mu_3 - \mu_3^2 = 0$ holds.

The terminating saddle points $(0,0)$, $(\mu_3, 0)$ and $(\mu_3, \frac{3}{2}(\mu_2 - 3\mu_3^2))$ are hyperbolic. The condition:

$$B_{g1}: \mu_1 + \mu_2\mu_3 - \mu_3^3 = 0$$

forms the global bifurcation set B_{g1}.

For $\mu_1 = 0$, B_{g1} reduces to the curves $\mu_3 = 0$ and $\mu_3 = \pm\sqrt{\mu_2}$ as depicted in Figure 3.25.

Going from region II (in (Figure 3.25) to region III a B_c-curve is crossed indicating cusp point bifurcation in the lower half plane; below this curve the center-saddle pair in the lower half plane has disappeared. Entering region IV from III, again a branch of B_c is passed on which a cusp in the upper half plane bifurcates into a center-saddle pair.

As a result the flow in region IV is characterized by an on-wall separation bubble together with a center-saddle pair above the wall.

On B_{g1} where $\mu_3 = 0$ and $\mu_2 > 0$ the flow is structurally unstable and consists of two counter-rotating vortices, surrounded each by three saddle connections in the upper half plane. Passing B_{g1} ($\mu_3 = 0$) to $\mu_3 < 0$ will reverse the mutual x-positions of the on-wall separation bubble and the center-saddle pair as they appear in region VI.

Proceeding now to the more general case $\mu_1 \neq 0$ we consider first the influence of μ_1 on the bifurcation sets and then on the phase portraits. For nonvanishing μ_1 we may use the parametrisation $\mu_2 = k(\mu_1/2)^{2/3}$, $\mu_3 = l(\mu_1/2)^{1/3}$ to obtain a unified picture of the bifurcation sets in a plane μ_1 = constant. Due to $\mu_1 \neq 0$ the bifurcation sets will become perturbed, as is shown in Figures 3.26 and 3.27.

The parametrized form of these sets becomes for $\mu_1 \neq 0$:

$$B_s : k = 3$$

$$B_c : \{98+63\,kl - 27\,l^3\}^2 = \{9l^2 + \frac{7}{3}k\}^3$$

$$B_{g1}: l^3 - kl - 2 = 0$$

Two special parameter combinations to be denoted by B_1 and B_2, appear at $(k,l) = (3,-1)$ and $[-3\sqrt[3]{\frac{7}{25}}, \frac{1}{3}\sqrt[3]{\frac{49}{5}}]$ respectively.

The bifurcation sets B_c and B_{g1} are tangent to B_s in B_1. Furthermore B_2 is a cusp point of B_c.

The bifurcation sets divide the parameter plane (μ_2,μ_3) in several domains (labelled I, I', II, II', III, III', IV, IV', V and VI) where topologically different phase portraits occur. These phase portraits are obtained by using standard methods such as local linearisation and singular-point analysis. Since these methods have been frequently discussed in previous paragraphs we omit detailed calculations here and we confine ourselves to stating the results.

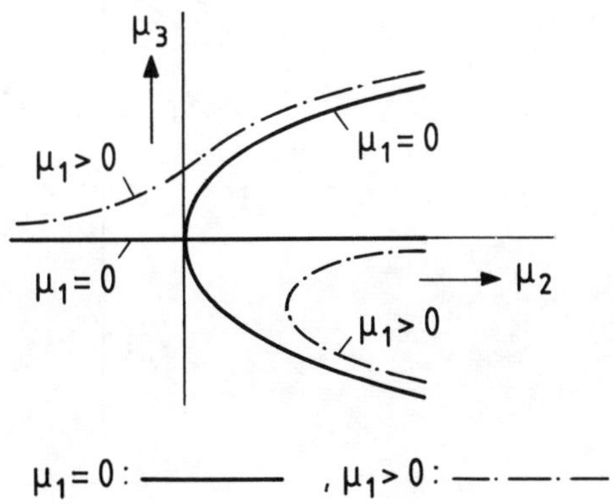

Fig. 3.26. Bifurcation set B_{g1}.

The influence of a nonvanishing μ_1 on the qualitative properties of B_s, B_c and B_{g1} is mainly concentrated around the μ_3-axis.
More precisely, beyond the interval $\mu_2(B_2) < \mu_2 < \mu_2(B_1)$ they approach, for increasing $|\mu_2|$, more closely to the form they have in the case $\mu_1 = 0$. This means that corresponding domains (equally labelled in Figure 3.27) at $\mu_1 = 0$ and $\mu_1 \neq 0$ will represent similar (topologically equivalent) phase portraits of Eq. (3.36). However, if μ_2 is sufficiently small, the influence of μ_1 is more severe and gives rise to the appearance of two new domains: V (near B_2) and VI (near B_1) for $\mu_1 \neq 0$.

The phase portraits in V and VI will be deduced from those found in adjacent domains and by passing the separating bifurcation set. In this way it follows that the flow in region VI is modelled by a hyperbolic saddle point on the wall, a center-saddle pair above the wall and a free saddle point in the lower half plane. For region V more discussion is needed.

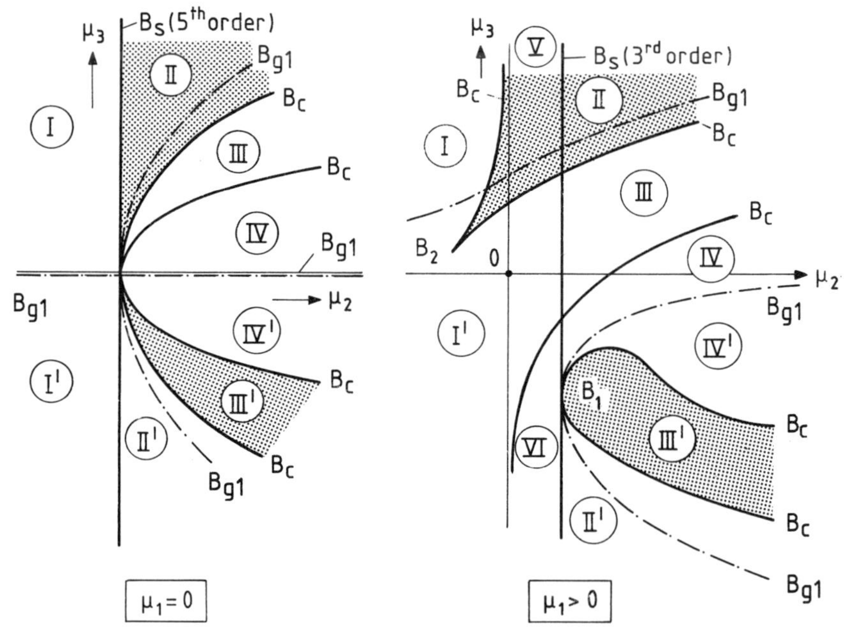

Fig. 3.27. Bifurcation sets B_s and B_c.

Region V is subdivided by B_{g1} into Va and Vb (for details see Figure 3.28). The flow in Va is obtained by applying a saddle point bifurcation in the flow in IIa. Then it is found that a hyperbolic saddle point, S_1, occurs on the wall, together with a center-saddle pair (C-S_2) and a hyperbolic saddle point, S_3, both in the lower half plane; streamlines entering the region between the center-saddle pair and the saddle point S_3 are prevented from crossing by the separatrix emanating from S_1.

The flow pattern in region Vb merits further consideration. We can obtain this pattern from the pattern in Va by crossing the global bifurcation set B_{g1}, the effect of which will

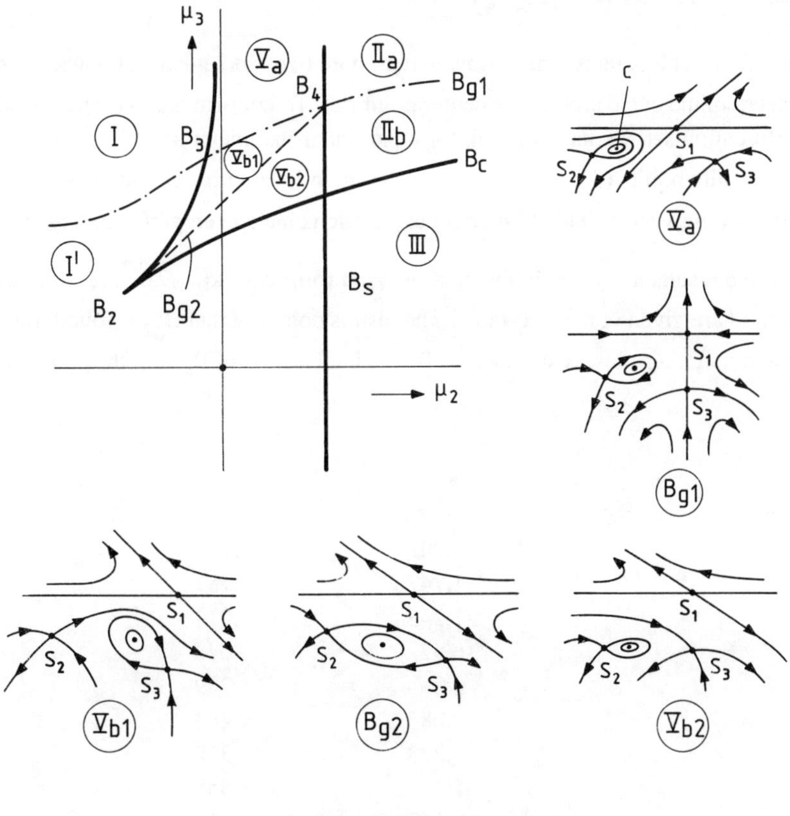

Fig. 3.28. Bifurcations near B_2.

be that (C-S_2) and S_3 will now be found on the same side of the separatrix through S_1. On B_{g1} there will now be a saddle-connection between S_1 and S_3, indicating structural instability in the global sense. This flow is obtained as well when going through the upper branch of B_c from I' to Vb. However, deriving the flow pattern from the pattern in IIb by crossing B_s (or alternatively from I', by crossing the lower branch of B_c), we obtain a pattern in which (C-S_3) forms a center-saddle pair, instead of (C-S_2). Consequently, two topologically different flow patterns seem to occur in region Vb. This discrepancy can be removed by assuming a second *global bifurcation set, called* B_{g2}, which divides Vb into the subdomains Vb1 and Vb2 such that the center-saddle pairs (C-S_3) and (C-S_2) occur in Vb1 and Vb2 respectively, see Figure 3.28.

The bifurcation set B_{g2} represents a flow pattern with two saddle-saddle connections forming a heteroclinic cycle around the center point (C). To confirm the existence of such a global bifurcation set, we have to look for a particular parameter set $\mu_3 = \mu_3(\mu_1,\mu_2)$ for which the saddle points (S_2) and (S_3) lie on the same level curve of the Hamiltonian of the corresponding vector field. Numerical calculations have been performed to approximate B_{g2}; the results are given in the parametrized form $\mu_2 = k(\mu_1/2)^{2/3}$, $\mu_3 = l(\mu_1/2)^{1/3}$ where k and l are given in the next table. The results point out that B_{g2} is found in region Vb only and that it terminates at B_2 and at B_4 the intersection of B_s with the upper branch of B_{g1}.

	k	l
B_2:	-1.963	.713
	-1.902	.729
	-1.747	.781
	-1.245	.914
	-6.57	1.078
	0.0	1.259
	.085	1.283
	.223	1.317
	1.217	1.558
	2.369	1.864
B_4:	3.000	2.000

Table: Numerical approximation of B_{g2}.

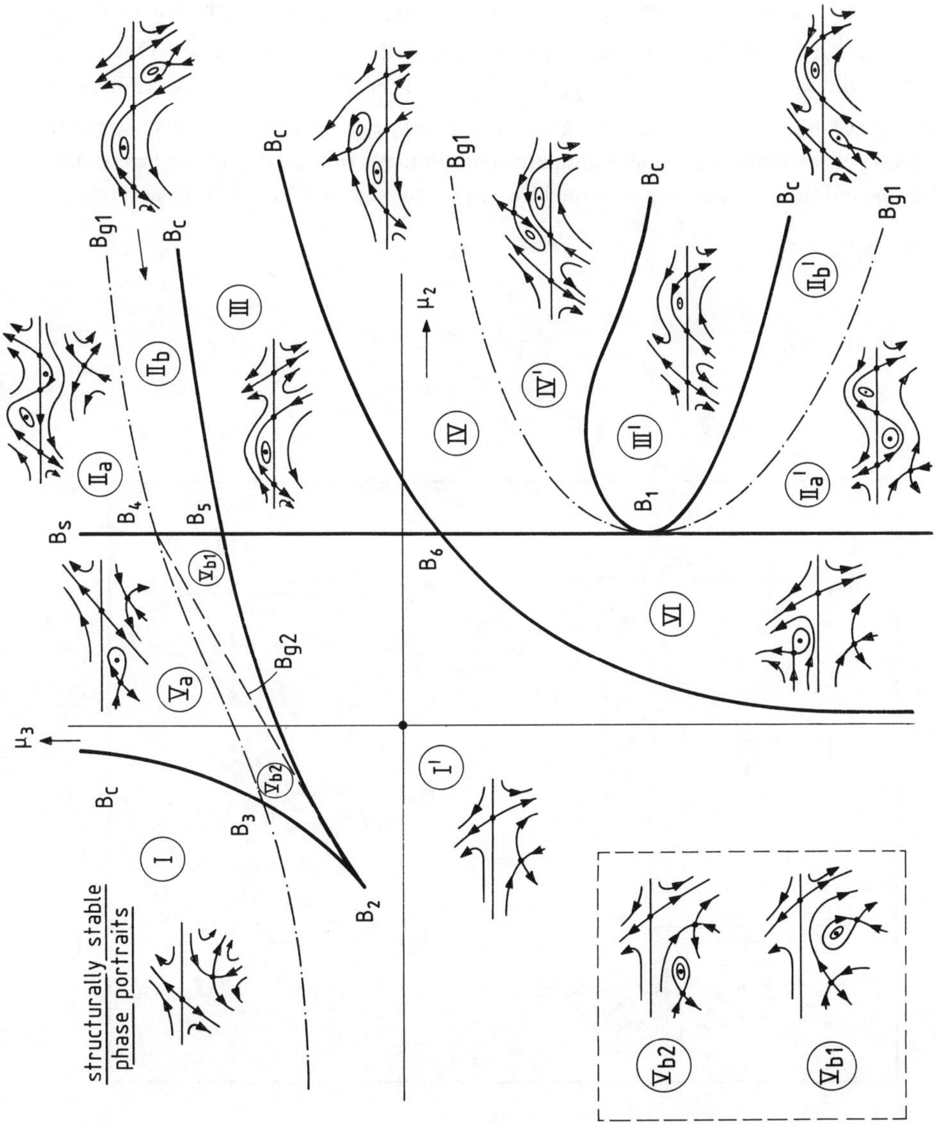

Fig. 3.29a. Unfolding of Eq. 3.36, $\mu_1 > 0$, bifurcation sets and structurally stable phase portraits.

All the previous results concerning $\mu_1 \neq 0$ may be brought together to obtain a total view of the physical unfolding of a higher-order singularity where the streamwise pressure gradient and the shear stress gradient, τ_{xx}, vanish simultaneously. The resulting unified picture is given in Figure 3.29 displaying the various properties (bifurcation sets and phase portraits) of the unfolding. Since parametrisation in terms of k and l appeared to be possible, the unfolding is actually governed by two parameters and has codimension two.

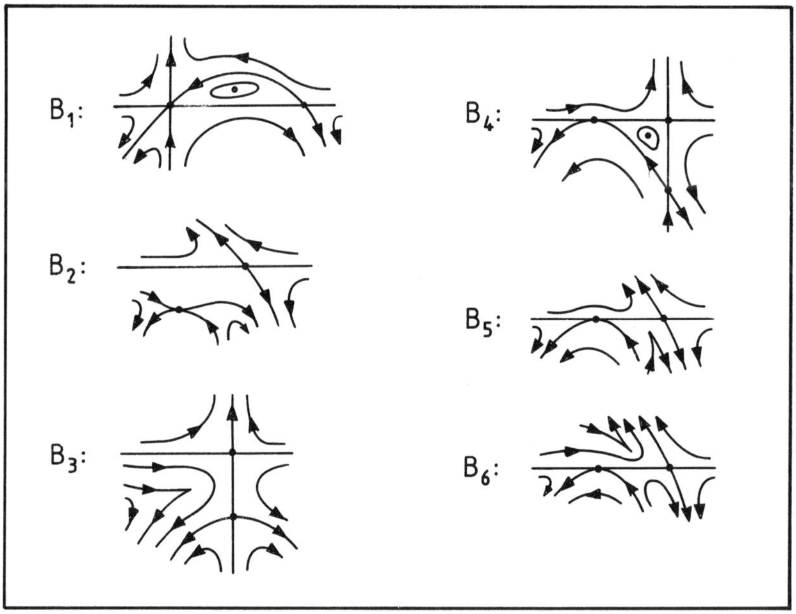

Fig. 3.29b. Codimension-two degenerate phase portraits of Eq. 3.36.

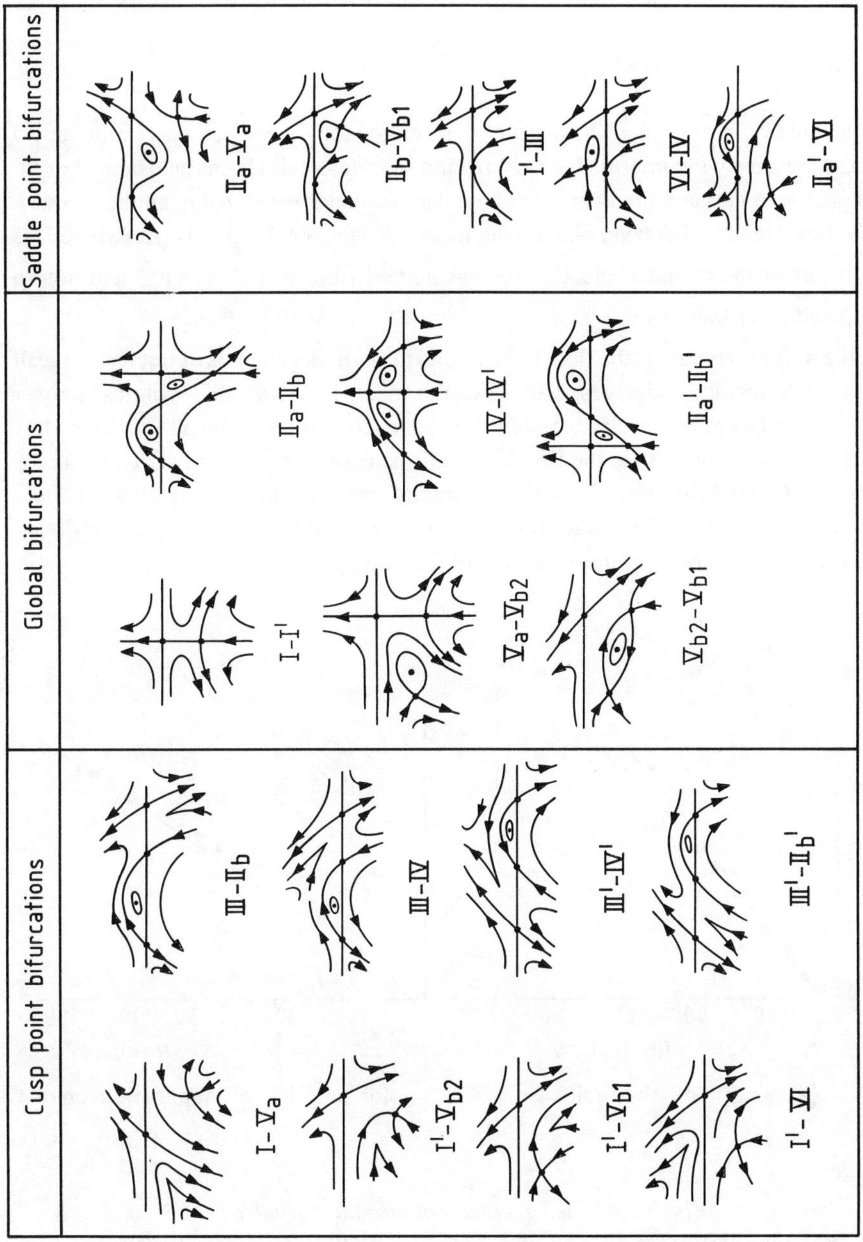

Fig. 3.29c. Codimension-one degenerate phase portraits of Eq. 3.36.

8. Viscous flow near a circular cylinder at low Reynolds numbers

8.1 Description of flow topology

Although the results presented above have been derived for flat walls, it may be attempted to discuss the steady incompressible flow behind a circular cylinder as an example of bifurcation at zero pressure gradient. We know that the nondimensional quantities, describing the flow field, will depend only on the Reynolds number $R = \rho.D.U_\infty/\mu$ (where D is the diameter of the cylinder and U_∞ the undisturbed free stream velocity) and not on D, U_∞ ρ and μ separately.

The value of R characterizes the flow field around the cylinder and it is of interest to recall how the flow topology changes with Reynolds number. At small Reynolds number ($R < 1$) inertia forces are negligible compared to viscous forces over most of the flow field. The dominant process in the flow is the diffusion of vorticity away from the body. At very low Reynolds number ($R << 1$) the vorticity spreads out fairly evenly in all directions so that the flow has fore- and-aft symmetry near the cylinder. This symmetry is reflected very well by Oseen's approximation of this flow.

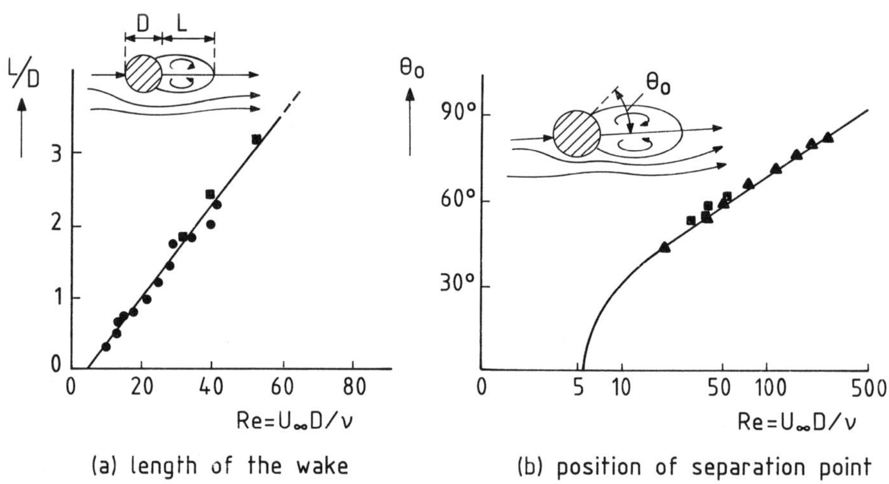

Fig. 3.30. Wake geometry of circular cylinder
● Taneda (1971) ■ Koromilas & Telionis (1980) ▲ Homann (1937)
——— theory: Pruppacher, Le Clair & Hamielec (1970)

At moderate Reynolds number, specified very roughly as $1 < R < 100$, inertia forces and viscous forces become comparable in magnitude and neither of them can be neglected. Vorticity diffused away from the body will be carried downstream and disturbs the aforementioned fore-and-aft symmetry, see Batchelor (1970). Furthermore there appears to be a definite Reynolds number, $R = R_c$, above which a region of slowly recirculating fluid occurs, forming a wake immediately behind the cylinder; if R is further increased this wake region becomes longer.

Taneda (1971) and Koromilas and Telionis (1980) have measured the length of the circulating region for different Reynolds numbers and their results suggest that this region first appears from $R = 5.2$ onwards, see Figure 3.30a. The experimental observations compare well with theoretical results as obtained from a numerical solution of the steady state NS-equations by Pruppacher, Le Clair and Hamielec (1970).

A similar conclusion may be drawn from Figure 3.30b which displays some numerical and experimental results concerning the location of the separation point on the cylinder surface for various Reynolds numbers.

8.2 Symmetrical bifurcations

The changing flow topology at a critical value of the Reynolds number will be reconsidered and seen as a local bifurcation occurring at the rear of the cylinder.

At the rear a second stagnation point appears and the flows over the upper and lower side meet and leave the surface of the cylinder. Due to symmetry the pressure gradient along the surface will be zero in the rear stagnation point.

The flow near this point can be described, at least approximately, by Eq. (3.10) which, for symmetrical flow conditions:

$$u(x,y) = -u(-x,y), \; v(x,y) = v(-x,y),$$

reduces to

$$\begin{aligned} \dot{x} &= a_2x + a_5xy + a_7x^3 - \frac{1}{2}a_7xy^2 + O(4) \\ \dot{y} &= -\frac{1}{2}a_2y - \frac{1}{3}a_5y^2 - \frac{3}{2}a_7x^2y + \frac{1}{2}a_7y^3 + O(4) \end{aligned} \tag{3.40}$$

The streamline $y = 0$ may be regarded as an approximation of the cylinder surface. Separation from this surface occurs at $(x,y) = (0,0)$ which is a hyperbolic saddle point if $a_2 \neq 0$.

Due to symmetry the shear stress and the pressure gradient along the wall will vanish there. For $a_2 = 0$ Eq. (3.40) shows a higher-order singularity at the separation point, indicating a structural instability of the flow pattern probably corresponding to the critical state $R=R_c$. Flow patterns occurring at near-critical Reynolds numbers may be found by unfolding the higher-order singularity ($a_5 = a_7 = -1$):

$$\dot{x} = -xy - x^3$$

$$\dot{y} = \frac{y^2}{3} + \frac{3}{2}x^2y$$

under the restriction of symmetrical flow conditions.
This means that $\mu_1 = \mu_3 = 0$ so that the number of bifurcation parameters is diminished and that the following one-parameter unfolding results

$$\dot{x} = \mu_2 x - xy - x^3$$

$$\dot{y} = -\frac{1}{2}\mu_2 y + \frac{y^2}{3} + \frac{3}{2}x^2y$$

where μ_2 denotes small perturbations of a_2 with respect to $a_2 = 0$.
Symmetrical bifurcations as caused by μ_2 are shown in Figure 3.31a.

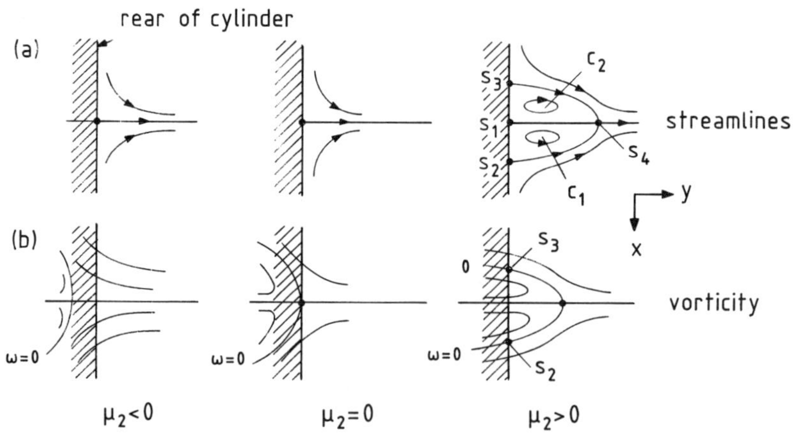

Fig. 3.31. Symmetrical bifurcations at the rear of the circular cylinder in steady flow.

Obviously the case $\mu_2 = 0$ reflects a critical state very similar to that occurring at $R = R_c$. Beyond the critical state ($\mu_2 > 0$) a cluster of singularities (saddles and centers) arises forming a closed recirculating region behind the cilinder; three stagnation points S_1: (0,0), $S_{2,3}$: $(\pm\sqrt{\mu_2}, 0)$ appear on the surface and three stagnation points appear in the flow field: S_4: $(0, \frac{3}{2}\mu_2)$ and $C_{1,2}(\pm\sqrt{\mu_2/7}, 6\mu_2/7)$, see Fig. 3.31a.

The relative positions of the saddle points S_2, S_3 and S_4 determine the general shape of the recirculating region. In the limit $\mu_2 \to 0$ this region shrinks in an 'oval' fashion such that its dimension normal to the cylinder (l) is smaller than along the cylinder surface (s). The ratio of the principal dimensions, $l/s = \frac{3}{2}\sqrt{\mu_2}$, may be expressed in local variables

$$\frac{l}{s} = \frac{-\sqrt{-6\tau_x \cdot \tau_{xxx}}}{2p_{xx}} \tag{3.41}$$

and will depend on the Reynolds number.

Thus unfolding theory leads to the conclusion that $l/s \to 0$ if $R \to R_c$ from post-critical values and that the bifurcation parameter μ_2 is related to the Reynolds number such that $\mu_2 \to 0$ if $R \to R_c$.

Another interesting aspect of the unfolding concerns the vorticity distribution $\omega(x,y)$ near the rear part of the cylinder. For symmetric bifurcations ($\mu_1 = \mu_3 = 0$), equation (3.40) can be used to derive the approximate vorticity distribution.

$$\omega(x,y) = -\mu_2 x + 2xy + x^3 + \frac{3}{2}xy^2 + O(4)$$

Lines of constant vorticity are displayed in Fig. 3.31b and yield the conclusion that the vorticity pattern near the rear stagnation point is mainly determined by the position of a saddle point on the axis of symmetry at $y = \frac{1}{2}\mu_2 + O(\mu_2^2)$.

This saddle point is found in the actual flow domain (wake region) for $\mu_2 > 0$ only. The level curve of zero vorticity: $\omega = 0$, has a parabolic shape, and intersects the surface of the cylinder in the separation points S_1 and S_3. The formation of the wake beyond the critical state $R = R_c$ concurs with the appearance of a saddle point in the vorticity pattern.

8.3 Asymmetrical bifurcations, transition scenario's

In this paragraph we consider asymmetric bifurcations of the higher-order singularity:

$$\dot{x} = -xy - x^3$$

$$\dot{y} = \frac{y^2}{3} + \frac{3}{2}x^2y$$

in order to investigate some characteristic properties of small asymmetric effects as they can appear if the flow near a circular cilinder is observed experimentally. Actually, deviations in the uniformity of the oncoming flow, imperfections of circular shape and intruding measurement techniques may introduce irregularities disturbing a perfectly symmetric flow pattern. Moreover, at higher Reynolds numbers unsteadiness and flow instability may appear so that symmetrical flow conditions, should they appear, cannot be maintained in time (Von Karman vortex street).

The influence of asymmetric disturbances on the steady flow near the critical state may be studied from the influence of the bifurcation parameters μ_1 and μ_3 on the physical unfolding:

$$\dot{x} = \mu_1 + \mu_2 x + \mu_3 y - xy - x^3$$

$$\dot{y} = -\frac{1}{2}\mu_2 y + \frac{y^2}{3} + \frac{3}{2}x^2y$$

Note that the bifurcation parameters μ_1 and μ_3 may be identified with the shear stress τ and the circumferential pressure gradient, respectively, at the rear of the cylinder.

The phase portraits and bifurcation sets of this unfolding are extensively studied in paragraph 3.7; a concise assembly of the results is provided in Figure 3.29. In order to apply these results to the flow at the rear of the cylinder we only need the phase portraits in the semiplane $y \geq 0$. The portraits at $y \geq 0$ together with the relevant bifurcation sets are reproduced in Figure 3.32 using the scaled variables $k=\mu_2[\frac{\mu_1}{2}]^{-2/3}$ and $l=\mu_3[\frac{\mu_1}{2}]^{-1/3}$ which are related to the circumferential gradients of shear stress τ_x and pressure p_x respectively. It shows a great variety of flow patterns which can occur at the rear of the cylinder and also various possibilities of sequences of flow patterns, when increasing the Reynolds number. In order to discuss the asymmetric bifurcations in a systematic way, we prefer to give therefore more attention to sequences of flow patterns than to the individual flow patterns. However, sequences can be chosen in various ways. Here we want to discuss those sequences which appear if the bifurcation parameter μ_2 (related to the circumferential shear stress gradient) increases whilst μ_1 and μ_3 are held constant. This idea is taken from paragraph 8.2 where a similar approach is followed to discuss the symmetrical bifurcations ($\mu_1 = 0,\ \mu_3 = 0$). Moreover, in that case an increase of μ_2 could be associated with an increase of the Reynolds number.

Consulting Figure 3.32 one observes that a continuous increase of μ_2, and also of k, causes the intersection of a sequel of bifurcation sets of different type (B_s, B_c, B_g). It involves the existence of a scenario of sequential flow patterns, linked together by bifurcations of structurally unstable flow situations.
The order in which consecutive bifurcations are passed through determines a particular scenario, moreover several scenario's seem to occur. Actually five different scenario's: S_1, S_2 ..., S_5 are distinguished in Figure 3.32.

All these scenario's reveal the common feature that asymmetric disturbances, as introduced by μ_1 and μ_3 are qualitatively unobtrusive as long as $\mu_2 < 0$, that is to say if the circumferential shear stress gradient τ_x remains negative. With respect to the symmetrical situation the flow pattern is slightly deformed in the sense that the separating streamline leaves the body obliquely dividing the wake flow into an 'obtuse' region and an 'acute' region.
If μ_2 is increased beyond $\mu_2 = 0$ the flow behind the cylinder is subjected to a sequence of bifurcations, making the influence of μ_1 and μ_3 more severe.
Depending on the order in which bifurcations of different kind succeed each other, various scenarios can be observed in Figure 3.32. Let us elaborate some details of these scenario's, for example those occurring in S_1 if μ_2 increases beyond $\mu_2 = 0$.

At $\mu_2 = 3\left[\frac{\mu_1}{2}\right]^{2/3}$ (k=3) a saddle point bifurcation is passed, announcing the formation of a separation bubble in the 'obtuse' part of the flow. Next a global bifurcation appears which turns the separating streamline counter-clockwise so that the bubble is now in the 'acute' region. Finally a cusp point bifurcation terminates the sequence and generates a closed flow domain with circulating flow between the separation bubble and the separating streamline.

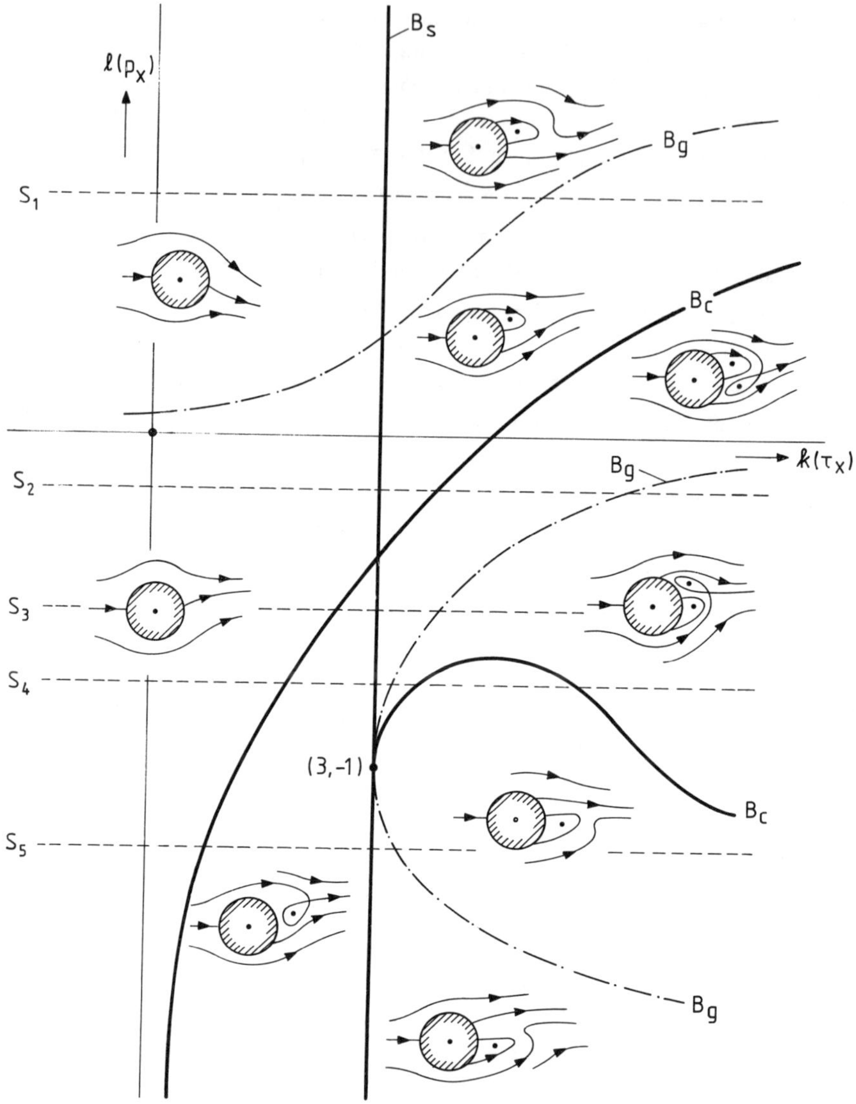

Fig. 3.32. Asymmetrical bifurcations at the rear of the circular cylinder in steady flow.

The resulting flow pattern is topologically different from that in a symmetrical situation where two counterrotating vortices form a closed domain with circulating flow behind the cylinder.
The occurrence of a free circulating region under steady flow conditions is essentially an asymmetric effect, signalling vortex shedding typical for unsteady quasi-periodic flows at higher Reynolds numbers.

Considering the scenario's S_2, S_3, S_4 and S_5 similar conclusions may be drawn from the possible sequences of flow patterns which can occur if μ_2 increases to postcritical values. A unified picture of these sequences (transition scenario's) is shown in Figure 3.33.

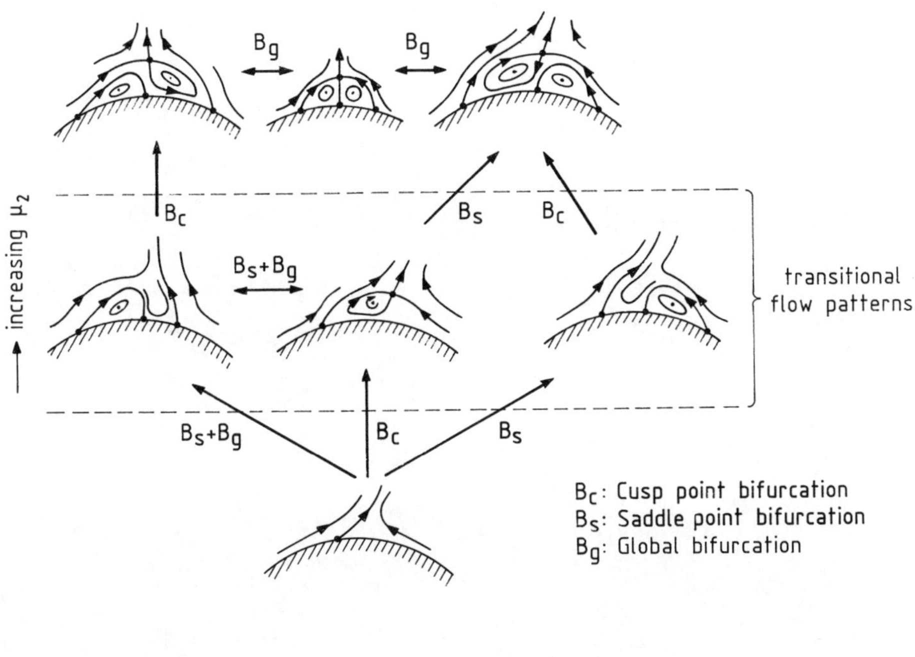

Fig. 3.33. Transition patterns in the steady flow of a circular cylinder.

On the basis of these results we conclude that asymmetrical effects introduce the possibility of more than one transition scenario. Moreover, such an asymmetrical transition includes several bifurcations suggesting that a certain range of Reynolds numbers has to be passed through before a fully developed wake region is established behind the cylinder. Lastly, the asymmetrical effects include the formation of a free region of circulating flow, reminding of an essential feature of vortex shedding, however without including the actual effect and unsteady nature of vortex shedding.

9. References

Batchelor, G.K. (1970)
An Introduction to fluid dynamics, Cambridge University Press, 255-263.

Danberg, J.E. and Fansler, K.S. (1975)
Separation-like similarity solutions on two-dimensional moving walls, AIAA-Journal, Vol. 13, no. 1; 110-112.

Dean, W.R. (1950)
Note on the motion of a liquid near a position of separation, Proc. Cambridge Phil. Soc. 46, 293-306.

Van Ingen, J.L. (1975)
On the calculation of laminar separatrix bubbles in two-dimensional incompressible flow, AGARD Conf. Proc., no. 168, Göttingen.

Inoue, D. (1981)
MRS criterion for flow separation over moving walls, AIAA-Journal, Vol. 19, no. 9, 1108-1111.

Koromilas, C.A. and Telionis, D.P. (1980)
Unsteady laminar separation: an experimental study, J. Fluid Mech., Vol. 97.

Legendre, R. (1955)
Decollement laminaire régulier, Comptes Rendus, Acad. Sci. Paris 241, 732-734.

Moore, F.K. (1958)
On the separation of the unsteady laminar boundary layers, in Boundary Layer Research (ed. H. Görtler), Springer Verlag Berlin, 296-310.

Oswatitsch, K. (1957)
Die Ablösungsbedingung von Grenzschichten, Symposium on boundary layer research, IUTAM Freiburg.

Pruppacher, H.R., Le Clair, B.P. and Hamielec, A.E. (1970)
Some relations between drag and flow pattern of viscous flow past a sphere and a cylinder at low and intermediate Reynolds number, J. Fluid Mech. Vol. 44.

Rott, N. (1956)
Unsteady viscous flow in the vicinity of a stagnation point, Quart. J. Appl. Mech., Vol. 13, 444-451.

Sears, W.R. (1956)
Some recent developments in airfoil theory, J. Aeronaut. Sci., Vol. 23, 490-499.

Sears, W.R. and Telionis, D.P. (1957)
Boundary layer separation in unsteady flow, SIAM, J. Appl. Math., Vol. 28, 215-235.

Takens, F. (1974)
Singularities of vector fields, Publ. Math. IHES, 43, 47-100.

Taneda, S. (1971)
Visualization experiments on unsteady vicous flows around cylinders and plates. In: Recent Research on Unsteady Boundary Layers (ed. E.A. Eichelbrenner), Vol. 2, Les Presses de l'Université Laval, Quebec.

Telionis, D.P. and Werle, M.J. (1973)
Boundary layer separation from downstream moving boundaries, J. Appl. Mech., Vol. 40, 369-374.

Tsahalis, D.T. and Telionis, D.P. (1973)
The effect of blowing on laminar separation, J. Appl. Mech., Vol. 40, 1133-1134.

Williams, J.C. III (1977)
Incompressible boundary layer separation, Annual Review of Fluid Mechanics (ed. M. van Dyke), Vol. 9, 113-144.

INDEX OF SUBJECTS

Nonlinear Topics in the Mathematical Sciences

An International Book Series dealing with Past, Current and Future Advances and Developments in the Mathematics of Nonlinear Science

1. M. S. Berger: *Mathematical Structures of Nonlinear Science.* An Introduction. 1990 ISBN 0-7923-0728-3

KLUWER ACADEMIC PUBLISHERS – DORDRECHT / BOSTON / LONDON